Sacrificing the Forest

Sacrificing the Forest

Environmental and
Social Struggles in Chiapas

Karen L. O'Brien

Westview
PRESS
A Member of the Perseus Books Group

Copyright © 1998 by Westview Press, A Member of the Perseus Books Group

Published in 1998 in the United States of America by Westview Press, 5500 Central Avenue, Boulder, Colorado 80301-2877, and in the United Kingdom by Westview Press, 12 Hid's Copse Road, Cumnor Hill, Oxford OX2 9JJ

Library of Congress Cataloging-in-Publication Data
O'Brien, Karen L.
  Sacrificing the forest : environmental and social struggles in
Chiapas / by Karen L. O'Brien.
      p. cm.
  Includes bibliographical references (p.   ) and index.
  ISBN 0-8133-6905-3 (hc) —ISBN 0-8133-3890-5 (pb)
    1. Deforestation—Mexico—Lacandona Forest.   2. Deforestation—
Social aspects—Mexico—Lacandona Forest.   3. Lacandona Forest
(Mexico).   I. Title.
SD418.3.M6037   1998
333.75'137'097275—dc21   97-47343
                                                                                                              CIP

PERSEUS
POD
ON DEMAND   10   9   8   7   6   5   4   3   2

Contents

Tables and Figures

Preface

The research for this book was undertaken as part of a larger study on deforestation and climate change in the Selva Lacandona. Originally, the analysis of deforestation was rather modest in scope, limited to evaluating deforestation around 24 climate stations scattered throughout the region. Much of this work was to be done through the interpretation of satellite images, verified by ground truth data collected in the field. However, as I traveled around the Selva Lacandona to visit the climate stations, it became increasingly evident that the patterns of deforestation existed in both historical and locational contexts. What had been generally presented as the obvious causes of deforestation in this region did not match the experiences and histories revealed to me during the course of my fieldwork.

Informal interviews conducted between 1991 and 1995 challenged me to broaden the scope of the deforestation analysis. Moving beyond a focus on the pixels of the satellite images, I began to explore the underlying causes of deforestation. It quickly became clear to me that deforestation was the outcome of a complex set of relations that extended well beyond the Selva Lacandona region. Over time, I concluded that Mexico's last tropical forests were being sacrificed, not to irrational land uses, but to economic and social realities.

The fieldwork also introduced me to the ecological realities of deforestation. Using the Chajul Tropical Biology Station as a base for my research on climate change, I was in close contact with biologists, ecologists, and environmentalists conducting various types of studies related to tropical ecology and conservation biology. Many of these people had years of experience in the Selva Lacandona, and were committed to finding solutions to deforestation that involved local communities. Others were clearly more interested in conducting their own research, and were annoyed at the encroachment of humans into their study areas. These interactions gave me a greater understanding of the struggles faced by conservationists, as well as insights on the politics involved with protecting the region's remaining tropical forests.

Over the course of nine visits to Chiapas and many flights over the Selva Lacandona region, I witnessed dramatic changes in the landscape. I also witnessed the emergence of a powerful social struggle that culminated in the Zapatista Uprising on January 1,1994. In this book, I attempt to make sense of both of these changes, and show how they are related, as well as how they are distinct.

This book does not attempt to cover all aspects of environmental and social struggles. For example, I have left out a discussion on much of the important work that has been done related to biodiversity in the Selva Lacandona, particularly the impacts of land use changes, illegal hunting, and trade in endangered species. This is, of course, one of the underlying reasons for concerns about deforestation. Although it is a critical component of conservation struggles in the Selva Lacandona region, I have tried not to spend too much time justifying conservation efforts. Instead, I have focused on the outcomes of these efforts and their role as "countervailing pressures" to deforestation.

Regarding social struggles, I have not discussed the political organizations that have formed in the Selva Lacandona region, including political affiliations, alliances, and the relationship of social movements to the Zapatista Uprising. This work has been sufficiently described in a number of publications. I have instead focused on social struggles in the broader context, namely the search for improved social welfare and expanded opportunities for economic participation. Likewise, I have not gone into detail on the origins and explanations for the Zapatista Uprising, except within the context of deforestation. I have tried to show that many of the economic, social, and political relations behind the uprising are identical to those forces responsible for much of the deforestation in the Selva Lacandona region.

In short, I have attempted to reconstruct and explain how the different patterns of deforestation in the Selva Lacandona have come about, and why the remaining forests take the form that they do. From a geographer's perspective, there is little mystery to the patterns — they are the manifestations of complex forces underlying both social and environmental struggles. What remains to be seen is whether these struggles will continue to develop along divergent paths, or whether they can be successfully fused to guarantee a future for the tropical forests of the Selva Lacandona.

Oslo, Norway
September, 1997

Acknowledgments

I would like to express my sincere thanks to the many people and institutions who helped me to carry out the research for this book. The initial research was funded through a Global Change Fellowship from NASA, as well as a Dissertation Improvement Grant from the National Science Foundation. Without such generous support, this research would not have been possible. I am grateful to Diana Liverman for her enthusiasm and advice, and to Donald Miller for encouraging me to publish part of my dissertation as a book.

I would also like to extend my gratitude to a number of people who helped me with my fieldwork and research in Chiapas. First, I'd like to thank Clementina Equíhua and Rodrigo Medellín for encouraging me to work in the Selva Lacandona region, and for their friendship and hospitality over the years. I would also like to thank Cecilia Conde, Rosa Ferrer, Blanca Herrera, Ramón Guerrero and family for their friendship and kindness. I was fortunate to have a number of great people assist me in my fieldwork at various times, including Kristian Stokke, Kim Steere, Heather Head, Erika Holden, Freddy Herrera, Diana Liverman, Robert Merideth, John Liverman, Elizabeth Wentz, Kimi Eisele, Charlene Floyd, and Dorothee Buerkle. I'd like to thank them all for their assistance and company. Finally, to the staff at the Chajul Tropical Biology Station, I would like to say *muchas gracias*.

I am grateful to Conservation International for providing me with logistical support on many occasions during my fieldwork. The Comisión Federal de Electricidad (CFE) and the Comisión Internacional de Limítes y Aguas (CILA) in Tuxtla Gutiérrez, Chiapas, also helped to facilitate my fieldwork. I would particularly like to thank Javier de la Maza, Jorge White, Luis Chavez, Victor Hugo, Ing. Gildardo Tipacamu Madrigal, Ing. Jacob Villatoro Velázquez, Ing. Carlos Santibañez Mata, Ing. Jorge Caballero Cordova, and Ing. Edgar Villalobos.

Finally, I would like to express my sincere thanks to Eduardo Iñigo and Ignacio March for their invaluable help throughout this project, including comments and suggestions on this manuscript. I would also like to thank Jane Barr for her continuous support, Nancy Hauth for her editorial expertise, and Karl Yambert and Jennifer Chen at Westview Press for being both helpful and patient. The warmest thanks go to my husband and son, Kristian and Jens Erik Stokke, for being so patient and understanding with me, and for encouraging me to see this book through to the end. My brothers, Tom and Ken O'Brien, also provided moral support throughout the research and writing phases. I would especially like to thank to my parents, Jim and Barbara O'Brien, for a lifetime of support and encouragement. I dedicate this book with love to my father, who showed such courage and strength in his own struggles with cancer.

To my father,
Jim O'Brien
(1935–1997)

1

Introduction

Environmental and Social Struggles in Chiapas

Since 1994, "Chiapas" has become embedded in the global lexicon as a reference to social inequality, indigenous resistance movements, and tropical rain forests. The Zapatista Uprising that began on January 1, 1994 channeled international attention toward the Selva Lacandona, or in the fanciful words of the media, to the "heart of the Lacandone jungle." This region, located in the eastern corner of Mexico's southern-most state, forms part of the largest remaining tropical rain forest in North America. Like most rain forests in Latin America, the Selva Lacandona has faced intense deforestation pressures since the 1960s. Although parts of the Selva Lacandona region are characterized by vast expanses of undisturbed tropical rain forest, many areas consist of mere fragments and remnants of a once-lush forest.

Prior to the Zapatista Uprising, the Selva Lacandona was the focus of national and international concern over the rapid rate of deforestation. While these concerns still prevail, over the past three years a growing faction within the international community has focused on social issues, particularly the marginalization of peasants and the possibilities for political mobilization at the grassroots level. For the socially-concerned, the tropical forest represents a formidable and somewhat romanticized terrain that is an impediment to economic development and greater social justice. For the environmentally-concerned, the Zapatista Uprising and its aftermath signifies an exacerbation of deforestation pressures, hastening the need to save the remaining forest. The distinction between those who would like to use the land of the Selva Lacandona and those who would like to conserve it represents one of the fundamental challenges of these times. Until this dichotomy is dissolved, the forest and the species that inhabit it will continue to disappear, leaving behind an increasingly impover-ished landscape.

A distinction between social and environmental struggles may be viewed as somewhat contrived, as many advocates of social change are indeed interested in protecting the forest for future generations, and many proponents of conservation recognize the importance of including local people in management plans. Nevertheless, there is evidence of growing friction in the Selva Lacandona region as peasants assert themelves through both organized and disorganized actions, calling for land distributions, improved services and infrastructure, autonomy for indigenous groups, and greater opportunities to participate in Mexico's economic development. Scientists, environmentalists, governmental institutions and nongovernmental organizations have at the same time mobilized to preserve what remains of the tropical rain forest, both within and outside of legally protected areas. Efforts to satisfy both concerns through the promotion of sustainable development schemes and community-oriented conservation projects have proved disappointing, and the Selva Lacandona remains a region embroiled in conflicts among diverse interest groups.

The social and environmental challenges confronting Chiapas today are actually closely related. Contemporary actions contributing to deforestation are embedded in a web of social, economic, and political relations with deep historical roots. In the case of the Selva Lacandona, the region has served as a focus of land speculation, a source of rapid capital accumulation, a refuge for displaced and disempowered people, a political safety valve for concessionary politics, a base for a revolutionary movement, and a spotlight for national and international conservation struggles. Many of the social, economic, and political processes driving deforestation have also contributed to widespread poverty and marginalization. The social conditions familiar to most parts of Chiapas impose further pressures on the remaining forest. It is fair to say that the Selva Lacandona is being sacrificed to social and economic realities.

This book will analyze the consistencies and contradictions between contemporary social and environmental struggles in Chiapas, focusing on the situation in the Selva Lacandona region. A political ecology approach will be used to show how the processes underlying deforestation are related to political, economic, and social relations, many of which are far removed from the forest. From this perspective, the future of the Selva Lacandona will be considered.

Three critical points are emphasized in this study. First, in order to understand the current situation in the Selva Lacandona, both the driving forces of deforestation *and* the countervailing pressures of conservation must be considered. Although it is essential to recognize that the spatial configuration of forested and deforested areas reflects a social history tied to the regional and international political economy, the same configuration also reflects the outcome of national and international conservation efforts. Over the past two decades, the politics of conservation have become particularly important in defining land use options in the Selva Lacandona region. The outcomes of these efforts have served to slow deforestation, impede it, or channel it to other areas. Yet these countervailing pressures have also created tensions in the region, which in some cases have erupted in conflicts. While conservation efforts aim to secure a certain amount of tropical forest for the future, they rarely address the underlying causes of deforestation. Consequently, they are unlikely to serve as a long-term solution to deforestation in the Selva Lacandona.

Second, it is critical to recognize that the driving forces of deforestation are played out very differently across the region. By invoking a geographical approach to an analysis of the political ecology of deforestation in the Selva Lacandona, it becomes clear that the complex web of relationships driving deforestation exists within locational contexts. Exploring these contexts contributes to an understanding of deforestation, the configuration of remaining forests, and possibilities for the future of the Selva Lacandona.

Third, although the links between social processes and environmental destruction are tight, those between social *changes* and tropical forest conservation are more tenuous. Many analyses that view tropical deforestation within a contextual framework related to social, economic, and political relations conclude that tropical forests can only be saved by coupling conservation efforts with movements for social and economic changes. Although this argument makes intuitive sense, the actual relationship between social change and tropical forest conservation is more complex, and deserving of a more careful analysis before conclusions can be drawn.

In the aftermath of the Zapatista Uprising, there have been recurrent calls for a new social and economic order in Mexico. Regionally, the changes proposed by the Zapatistas would not necessarily benefit the remaining forests of the Selva Lacandona. Instead, they suggest a greater integration of the region into the national and international market economies and increased land use pressures. Despite the fact that environmental and social struggles emanate from similar roots in the Selva Lacandona region, the final goals are quite distinct. Unless the goals can be fused, the future of the Selva Lacandona will continue to be shaped by the tensions between social, economic, and environmental objectives.

Resources and Realities

Chiapas, the southernmost state of Mexico, has been succinctly described by Benjamin (1989) as "a rich land, a poor people" (Table 1.1). This description is fitting because Chiapas is a state with extensive forests, rich agricultural lands, abundant water, significant oil reserves, and plentiful coastal resources along the Pacific Ocean. At the same time, social and economic indicators suggest that the people of Chiapas are among the poorest in Mexico. An estimated one-third of the population is malnourished; 60 percent of school-age children do not have access to education; almost 20 percent of the population lives in one-room dwellings; and 30 percent of the homes lacked electricity in 1990. There is only one doctor for every 1,500 - 3,500 people and the per capita income is about $1,466, compared to $3,000 for all of Mexico (Monroy 1994). Moreover, most of the wealth within the state is highly concentrated, with an estimated 20 families owning the best land (García de León 1984; Benjamin 1989). The discrepancies can be attributed to almost two centuries of feudal production relations and government policies dictated by landowners, merchants, and professionals who used the government to promote personal interests, often in the name of modernization (Benjamin 1989).

Currently, the population of Chiapas is estimated to be 3.7 million. The growth rate between 1980 and 1990 was about 4.5 percent, which leads to a projected population of 5 million by the year 2000 (Monroy 1994). A majority of the population is of indigenous background, and over 75 percent speak a non-Spanish dialect. The population is concentrated in 18,476 cities and villages spread throughout 7,500 km^2 of countryside, equivalent to 3.8 percent of Mexico's total land area (INEGI 1990). Chiapas is divided into 111 regional administrative units, known as *municipios*. Three of these municipios (Ocosingo, Las Margaritas, and Altamirano) contain a large part of the forests or forest remnants of the Selva Lacandona. Parts of the municipios of Palenque, Chilón, La Trinitaria and La Independencia also coincide with the Selva Lacandona region.[1]

The tropical forest ecosystems of the Selva Lacandona form part of the largest remaining tropical rain forest in North America. The Selva Lacandona is located in the Usumacinta river basin of eastern Chiapas, contiguous with the still-forested Petén region of Guatemala and the semi-deciduous forests of the Yucatan peninsula (Figure 1.1). The Usumacinta basin is the largest in Mexico, receiving a significant part of the country's total rainfall and providing immense hydroelectric potential. The terrain of this region consists of parallel northwest to

TABLE 1.1 Statistical overview of Chiapas.

A Rich Land ...		A Poor People [1]	
Land area	75,210 km^2	Population in 1995	3,707,135
Area of forests	28,812 km^2	Annual growth rate of population (percent)	4.51
Water resources (percent of Mexico's total water supply)	30	Percent of population under 15 years	44
Value of agricultural production (thousands of new pesbs)	2,330,199.3	Percent of population above 15 economically active	45
Value of livestock production (thousands of new pesos)	6,544,651.0	Per capita income (U.S. dollars)	1,466
Production of hydroelectric energy (thousands of KwH)	9,344,922.0	No. of medical personnel per 1,000 people	0.68
Production of crude oil (thousands of barrels)	20,703.7	Percent of illiteracy in population above 15 yrs.	30
Production of natural gas (million cubic feet)	172,097.5	Percent of population without running water in 1995	43
Tourism receipts (thousands of new pesos)	985,874	Percent of population without electricity	35

All data are from 1992, unless otherwise noted.
1 The description of Chiapas as "A Rich Land, A Poor People" is borrowed from Benjamin, 1989.
Sources: INEGI 1993; Monroy 1994; Comisión Nacional de Agual 1996; CONAPO 1997.

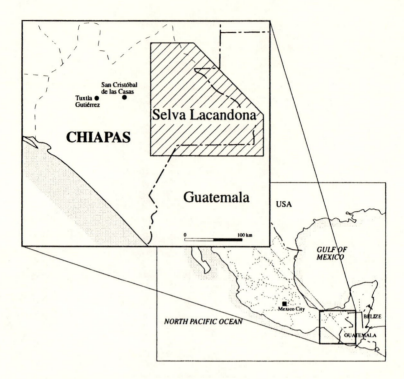

FIGURE 1.1   The Selva Lacandona region of Chiapas, Mexico.

southeast mountain chains which gradually decrease in elevation from
1800 meters above sea level in the western part of the region to close to
sea level in the southeastern corner. The vegetation of the region is
dominated by lowland and montane rain forests, as well as pine-oak
forests at higher elevations.

The Selva Lacandona is home to an enormous diversity of plants and
animals. This region alone contains about 25 percent of Mexico's total
species diversity, including at least 112 species of non-marine
mammals, 345 species of birds, 800 species of butterflies, and 4,000
species of vascular plants (Medellín 1994). It is considered to be an
important center for North American biodiversity, but also one of the
most endangered.

It has been claimed that as much as two-thirds of an original
1,500,000 hectares of forest has been cut and converted to pasture or
cropland over the past 40 years (Fuentes Aguilar and Soto Mora 1992).
This has raised great concern among scientists and environmentalists,
both nationally and internationally. In 1978, 331,200 hectares of the

Selva Lacandona forest was protected as part of the Montes Azules Biosphere Reserve (Diario Oficial 1978). This area was supplemented with the creation of several smaller protected areas in 1992. Nevertheless, the land both inside and outside of the reserves has been subject to continuing deforestation.

Deforestation as a Global Issue

The Selva Lacandona has attracted international attention for over a decade, as tropical deforestation has been increasingly viewed as a global issue. While deforestation in temperate countries is considered to be a serious problem, internationally there has been a much stronger emphasis on the loss of tropical forests, particularly rain forests (Williams 1994). Concerns related to biodiversity loss and climate change have underscored the importance of tropical forests to the global environment. These concerns have led some to suggest that tropical forests be considered a part of the global commons. Such proposals have provoked strong criticisms from tropical countries, which convincingly argue that forests should remain under national sovereignty (Shiva 1993).

In framing tropical deforestation as a global problem, the international community has expended great efforts in establishing the causes, along with a litany of potential solutions. The direct causes of deforestation usually include logging, conversion of forest to agriculture or pasture lands, fuelwood extraction, population growth, and the expansion of human settlements. It has also been acknowledged that the causes of tropical deforestation are tied to actions in temperate countries, through international trade in timber and beef, external debt, and multilateral bank policies. The proposed solutions attempt to address these causes, and include logging bans, the introduction of sustainable forestry and agroforestry techniques, programs in family planning, the creation of biosphere reserves or national parks, marketing of rain forest products, boycotts of tropical timber and beef, and debt-for-nature swaps. Along similar lines, many have called for local populations to end irrational land use practices and pursue "rational" management of natural resources.

Yet deforestation continues, leading many to believe that solutions to halt irrational land use practices are not being applied fast enough, for lack of financial resources or political will. Although this interpretation is pervasive in many circles, it has also been increasingly challenged by researchers who have examined deforestation in relation to the local, regional, and global political economy. Using an approach often referred to as "political ecology," these studies go beyond listing and describing the causes of deforestation. Instead, they seek to explain how the processes underlying deforestation are related to political, economic, and social relations. Such analyses show that tropical deforestation is in most cases a rational response to forces which span from the local to the global in scale.

A Paradigm Shift

Explanations for deforestation vary tremendously, depending on the depth and scope of the analysis. Perusing the expanding body of literature published on deforestation over the past 20 years, one senses that a change is underway in terms of how researchers interpret the causes of deforestation. This shift, alluded to above, involves a move away from actor- or agency-oriented explanations for deforestation, and toward a more critical perspective that emphasizes the underlying or driving forces. Deforestation is being increasingly interpreted within the context of social, economic, and political relations. There is a growing recognition that there exist both direct or proximate causes of deforestation, as well as underlying causes or driving forces.

A deeper examination of the literature suggests that this change is more than mere revisionism. In fact, changes in the ways that the causes of deforestation are portrayed increasingly resemble a paradigm shift. Evidence of this shift is not limited to one type of analysis, but rather can be found in deforestation research ranging from individual case studies to global regression models. Although this shift may in part be attributed to theoretical developments in individual disciplines, it largely reflects the growing number of scholars involved in deforestation research who have found it necessary to challenge some of the unsatisfying conclusions of earlier studies.

The earliest reports on the causes of tropical deforestation focused primarily upon commercial exploitation, particularly the logging of tropical hardwoods. Beginning in about 1980, studies of tropical deforestation turned attention toward the different agents of destruction (Allen and Barnes 1985). For example, Myers (1980; 1984) implicated the commercial timber harvester, the fuelwood gatherer, the forest farmer, and, in some parts of the world, the cattle rancher in the decline of tropical moist forests. The driving forces behind these agents were population growth and a concomitant increase in the number of shifting cultivators, tropical wood consumption, and international demand for beef. Rural poverty was also seen as a fundamental cause of deforestation. For a number of years, these causes of deforestation received tremendous international attention. However, the ensuing analyses did not provide satisfactory explanations for the continuing loss of forests. By the early 1990s, there was a growing consensus that the roots of the deforestation problem lie much deeper than the common understanding implied. The consensus was reflected in Myers' call for a "fundamental shift in our understanding of tropical forests and their part in the human enterprise" (Myers 1992:22).

The literature that has emerged over the past decade or so has put much greater emphasis on the driving forces of deforestation. Such studies include not only descriptions of the causes, but also critical analyses. For example, Gillis and Repetto (1988) emphasized the role of government policies in driving deforestation. Their analysis demonstrated that the issue of deforestation was much more complex than a question of economic development versus resource conservation, as most of the development policies proved to be unquestioned failures anyway. Taking an historical perspective, Richards and Tucker (1988) showed that the transformation of natural systems has been intimately tied to the transformation of human systems. They contended that a broad range of issues must be considered to fully understand the issue of deforestation.

Focusing on deforestation in the Amazon, Bunker (1985) turned from looking to theory for explanations of the continued underdevelopment in the region, to looking at the social, economic, and environmental history of the region. He concluded that the relationship between humans and nature in the Amazon can be seen historically as the result of the subordination of both to progressively wider and more complex economic and political systems.

Expanding on these themes and explicitly developing a political ecology approach to understanding deforestation in the Amazon, Schmink and Wood (1987) addressed the question of why so many development projects and conservation efforts for the Amazon have gone awry. They stressed the contradiction between conservation and long-term economic growth, with its goals of expanded production and short-term accumulation. Emphasizing the changing socioeconomic and political situations that shape the perceptions and behavior of individuals, they showed that each different actor or social group maintains a "rationality" for forest use, and that the multiple users are often in conflict (Schmink and Wood 1987; Schmink 1994). Hecht and Cockburn (1989) also argued that deforestation in the Amazon is a logical outcome of the region's historic and economic history, culminating in struggles related to justice and distribution.

In Central America, Schwartz (1995) demonstrated that tree removal is a result of rational decision-making in the Petén region of Guatemala, driven in part by macro-socioeconomic and political processes. To explain deforestation in southern Honduras, Stonich and DeWalt (1996) focused on the interconnections between deforestation and the export-led development model, the ongoing economic crisis, policies and actions of the state, competition among various classes and interest groups, and survival strategies of an increasingly impoverished rural population.

In their study of deforestation in Costa Rica, Vandermeer and Perfecto (1995) presented political ecology as an alternative philosophical approach that emphasized issues of both land and food security. They stressed that the true causes of deforestation are far more complicated than the proximate factors that are so frequently cited. In fact, they argued that "the nature of the complications *is* the cause" (Vandermeer and Perfecto 1995:xi, italics in original).

Deforestation has thus been increasingly considered a land use change that is intricately tied to social and economic relations, political power struggles, and changes in the international political economy. However, it is important to note that the alleged paradigm shift is neither categorical, nor does it adhere to a strict temporal scale. Some early studies focused on the complex relations that form the root causes of deforestation, and some recent studies continue to place emphasis on direct manifestations or proximate causes of deforestation, invoking neo-Malthusian paradigms, neoclassical economics, or dependency theories. Stonich (1993) presents a critique of these competing paradigms, and argues that although they may identify one or more factors contributing to deforestation in tropical countries, none of the models can provide a comprehensive explanation for deforestation.

The existence of competing paradigms draws out an important point worth emphasizing: The paradigm shift discussed above does not represent a move from groundless, inaccurate explanations toward a correct understanding or "truth." Instead, it reflects an emerging trend toward a more comprehensive understanding of the complexity of interactions driving deforestation. A political ecology approach tends to show that there is seldom one "main underlying cause" that can explain deforestation. Instead, a web of structural relationships emerges that explains or rationalizes what are perceived by many to be irrational land use decisions.

Overview

The purpose of this book is to examine environmental and social struggles in the Selva Lacandona, and the ways in which they impact deforestation. Using a political ecology approach, both the driving forces of deforestation and the countervailing pressures of conservation will be analyzed. From this analysis, it becomes evident that the marked distinction between the goals of conservationists and the goals of those seeking to improve the welfare and economic situation of individuals or local communities is likely to lead to the continued destruction of the forest.

To set the framework for this study, a political ecology approach is discussed in Chapter 2 and presented as a means of understanding the complex set of relations driving deforestation. The proximate causes of deforestation in the Selva Lacandona are then presented, followed by an abridged analysis based on a political ecology approach. This synopsis emphasizes some of the points mentioned above, and paves the way for a more detailed analysis of deforestation in the Selva Lacandona.

The ways in which the Selva Lacandona has been reconfigured over the past three decades are discussed in Chapter 3. The region is described in its historical context, and some of the most important transformations are outlined. Frequently cited rates of deforestation are presented, along with an original estimate based on the interpretation of satellite imagery. Next, the spatial and temporal heterogeneity of deforestation is emphasized by focusing on five subregions: the Zona Norte; Las Cañadas of Ocosingo; Las Margaritas; Marqués de Comillas; and the Montes Azules Biosphere Reserve. This analysis serves as a point of departure for examining the environmental and social struggles in Chiapas today.

To understand how the current configuration of the Selva Lacandona came about, one can begin with the role of extractive industries. Chapter 4 describes the role that logging and petroleum exploration have played in terms of both the economy and ecology of the region, as well as the way that those roles have changed over time. Although these industries are commonly considered responsible for much of the deforestation in the Selva Lacandona, their impacts have varied over both time and space. As a production activity with the potential to benefit local economies, the development of community forestry within the Selva Lacandona has been impeded by a state monopoly on logging and by government policies aimed at protecting the forest. In relative terms, extractive industries have benefitted very few, while facilitating access to the forest for many. In fact, the indirect role of extractive industries has been far more important than the direct role, largely through the construction of roads and infrastructure.

The existence of roads into the forest does not lead to deforestation in and of itself. Once such apertures exist, the driving forces behind deforestation become important. In Chapter 5, agricultural transformations in Mexico are discussed, along with the changes or lack of changes in agrarian relations in Chiapas. These changes provide a basis for understanding the forces behind colonization in the Selva Lacandona, as well as the land use patterns that have emerged within the forest. Within the context of agricultural transformations and the agrarian politics of Chiapas, the expansion of the agricultural frontier into the Selva Lacandona can hardly be called an "irrational" form of land use.

In Chapter 6, the colonization process is described, along with the settlement patterns that have resulted. The agricultural systems adopted by the colonizers are also considered, including the rationale behind the expansion of cattle ranching among both small-scale and large-scale farmers. The idea that colonizers aim for self-sufficiency and are completely removed from the market economy is reexamined. Most do, in fact, seek to produce a surplus for the market as rapidly as possible. Once settled in the forest, peasants struggle to make a living and establish a role in the rapidly changing global economy.

The role of peasants in the global economy came to the forefront of international debates in the aftermath of the Zapatista Uprising of January, 1994. The direction of Mexico's economic policies was questioned, and the implications for the rural population were reconsidered. This political upheaval was the second in little over a decade to impact the Selva Lacandona region. The first was the arrival of thousands of refugees from Guatemala during the early 1980s.

The ways that these two political upheavals have affected the region are considered in Chapter 7. The underlying causes of the guerrilla war in Guatemala are briefly discussed, along with the consequences for the economy and ecology of the border region of the Selva Lacandona. Similarly, the background to the Zapatista Uprising is presented and related to the driving forces of deforestation discussed in the previous chapters. The immediate implications of the uprising are explored, including increased militarization of the area and the rapid development of roads and infrastructure in select areas of the Selva Lacandona. A surge of land invasions in protected areas can also be tied indirectly to the Zapatista Uprising and the uncertain political climate in the region.

These protected areas are the core achievement of conservation efforts in the Selva Lacandona region. In addition to the declaration of protected areas, a series of international debt-for-nature swaps, sustainable development projects, and international funds for conservation have fortified efforts to protect the Selva Lacandona. The ways that these efforts have served as countervailing pressures on deforestation are discussed in Chapter 8. Although the conservation community has developed political savvy and a greater sophistication in the management and marketing of the Selva Lacandona, conflicts over land tenure and the role of logging in the Selva Lacandona region can be cited as two examples of the growing rift between environmental and social struggles in the Selva Lacandona.

In Chapter 9, questions regarding the future of the Selva Lacandona are addressed. It is emphasized that the tropical forests of the Selva Lacandona have been sacrificed to economic and social realities. These realities are shaped by land and labor relations forged by state politics in Chiapas, land use regulations established by the Mexican government, and the integration of Chiapas into the global economy. The sacrifice has occurred to the distress of the conservation community, which has expended enormous resources and personal energy to contain the destruction of Mexico's largest remaining tropical rain forest.

The point to stress here is that the contemporary situation in the Selva Lacandona is marked by two distinct struggles which are increasingly at odds with one another. Environmental struggles to protect the remaining tropical rain forest and social struggles to improve the lives of a highly marginalized population are creating tensions and conflicts that can only hasten the loss of forest.

Yet social changes alone are not enough to secure the future of the forest. There is little guarantee that social changes, particularly those sought by the Zapatistas, will not reproduce the same relations that have led to the destruction of a large part of the forest. The future of the forest must be explicitly considered when pursuing social changes that will undoubtedly impact the economy and ecology of the region.

Unresolved Issues

One of the most important unresolved issues regarding the future of the Selva Lacandona lies in the settlement of the Zapatista conflict. Whether it be negotiated or non-negotiated, the outcome of this social movement is likely to be decisive in securing the forest or condemning it to history.

Reading through the communiqués of the *Ejército Zapatista de Liberación Nacional* (EZLN) and negotiations with government representatives, it becomes quite clear that deforestation is not an issue on the Zapatista's agenda. On the rare occasions that natural resources are mentioned, it is in reference to ending the plunder by outsiders, particularly with respect to petroleum and hydroelectric power. Taking the Zapatista's demands into consideration, one foresees the future Selva Lacandona as a region fully integrated and developed, replete with infrastructure, services, and amenities. While this vision may be desirable from a social perspective, environmental struggles would most likely be lost.

If the Mexican government's current strategy of heavy militarization of the Selva Lacandona region continues, then the future of the forest is likely to be no better off. Many feel that the military will maintain the low intensity warfare that began in 1995, in an attempt take control of the situation and slowly destroy the resistance, putting an unceremonious end to the Zapatista conflict. The occupation of a large part of the Selva Lacandona with thousands of military personnel and armored vehicles has already altered social and economic relations in the region, and will undoubtedly lead to further disruptions. The mobilization of the military within the confines of the Montes Azules Biosphere Reserve suggests that the future of the Selva Lacandona is indeed precarious.

From all angles, it appears that the Selva Lacandona at the turn of the 21st century will be a vastly different place from the Selva Lacandona of thirty, fifty, or one hundred years ago. Given the nature of the driving forces of deforestation that are explored in this book, it is unlikely that environmental struggles alone will save the tropical forest. Likewise, social struggles alone are apt to result in continued deforestation. Unless the two struggles can develop a common ground, tropical forests will continue to be sacrificed to the realities of the day.

2

The Political Ecology
of Deforestation

Political Ecology

Widespread concern over tropical deforestation has resulted in hundreds of case studies, reports, models, and policy analyses that aim to identify the causes of deforestation. As discussed in the previous chapter, there has been a gradual paradigm shift in deforestation studies over the past two decades. In a broad sense, the paradigm shift represents a growing recognition that the causes of deforestation are more complex than they outwardly appear. This can be seen in subtle changes, such as the replacement of the term "shifting cultivators" as agents of destruction with "shifted cultivators" (compare, for example, Myers, 1980 and 1992). The phrase "shifted cultivators" implies that there is more to the story regarding why these cultivators are in tropical forests, and it raises the question: *Why were they shifted?* The theoretical approach described here as political ecology sees the latter question as essential to understanding tropical deforestation.

The term political ecology is rather vaguely defined, and has been used to refer to several quite distinct frameworks of analysis. The term has been appropriated by some to describe any number of studies related to the environment (Atkinson 1991) and by others to represent a critique of contemporary human-environment relationships (Lipietz 1995). Other interpretations have emerged from a variety of disciplines and subdisciplines. For example, Bryant (1992) argued that political ecology represents an emerging research agenda in Third World studies, as a move toward an integrated understanding of how environmental and political forces interact to mediate social and environmental change. Although this definition takes a largely political perspective, it recognizes that political ecology is "inclusive and sensitive to the interplay of diverse socio-political forces, and the relationship of those forces to environmental change" (Bryant 1992:14).

Appropriate to the understanding of land degradation and tropical deforestation, a political ecology approach has developed largely within the fields of geography and anthropology. The perspective emerged from a number of directions, including human and cultural ecology and political economy. More important, it has developed its identity through a collection of studies, rather than as a formal framework of analysis. For this reason, the term remains somewhat amorphous. Blaikie and Brookfield (1987:17) are usually credited with providing a general definition of political ecology, whereby it "combines the concerns of ecology and a broadly defined political economy." Blaikie and Brookfield's regional political ecology approach follows a chain of explanation, starting with land managers and their direct relations with the land, then linking to their relations with each other and with groups in the wider society, and finally to the state and world economy. The contribution of Blaikie and Brookfield to political ecology lies not in their definition of the term, nor in the chain of explanation that they elaborated, but rather in their attempt "to build a theory that allows for complexity, and identifies the sources of that complexity" (Blaikie and Brookfield 1987:239).

A more recent attempt to develop the framework of political ecology was presented by Greenberg and Park (1994). They argued that although political ecology has a broad interdisciplinary emphasis, two major theoretical thrusts can be identified: "These are political economy, with its insistence on the need to link the distribution of power with productive activity and ecological analysis, with its broader vision of bio-environmental relationships" (Greenberg and Park 1994:1). Political ecology is considered an historical outgrowth of the central questions asked by the social sciences about relationships between humans and nature in response to a developing consensus that it is "not enough to focus on local cultural dynamics or international exchange relations, and that the past and present policy, politics or political economy in general and the environment needs to be explicitly addressed" (Greenberg and Park 1994:8). Although they elaborated on the theoretical underpinnings of political ecology, Greenberg and Park insist that it is ill-advised to define the term. Instead, they contend that all legitimate forms of political ecology will share some similarities, but they need not share a common core.

Political ecology thus offers a flexible and dynamic framework for investigating the underlying causes of tropical deforestation. There are no set formulas for studies based on political ecology, and the framework can be readily adapted to the circumstances in each particular study area. This lack of formalism is one of the strengths of the political ecology approach. In contrast to the one-dimensional lists of direct causes or underlying forces that characterize much of the deforestation literature, a political ecology approach recognizes the inherent complexity of the driving forces. Invoking a geographical perspective to political ecology, it becomes possible to distinguish patterns and trends within the region, leading to a more comprehensive understanding of how deforestation is manifested across space and time.

The Proximate Causes

The Selva Lacandona is Mexico's largest remaining tropical rain forest, and forms an important center for Mexican and global biodiversity. The region covers an area of approximately 1.9 million hectares in the eastern part of Chiapas, directly west of the Usumacinta river and north of the Mexico-Guatemala border. The tropical forest vegetation of the region has been substantially altered by human impacts, with deforestation becoming increasingly significant over the past three decades. Some estimate that as much as two-thirds or three-quarters of the tropical forest has been replaced by agricultural crops, pasture, or secondary growth (Hernández 1990; Fuentes Aguilar and Soto Mora 1992). However, the analysis of satellite imagery presented in Chapter 3 suggests that those numbers may be exaggerated depictions of regional deforestation. While some areas within the Selva Lacandona have been severely deforested, other areas remain largely intact. The analysis does conclude that an estimated 40 percent of the forest had been destroyed by the early 1990s, with most of this concentrated in the valleys in the northern and western parts of the region. The largest areas of remaining forest were located in the southeast, particularly within the Montes Azules Biosphere Reserve and in the Marqués de Comillas subregion.

The causes of deforestation in the Selva Lacandona have been widely discussed in both the media and academic literature (Ovilla and Díaz López 1985; Rojas 1986; Dichtl 1987; *El Día* 1987; de Vos 1988b; Lazcano-Barrero et al. 1992). Conservation biologists and the media tend to portray the current situation in terms of an environmental catastrophe, whereas social scientists have focused more on anthropological, sociological, historical, or economic perspectives of various processes leading to deforestation. To date, there have been no comprehensive geographical studies that provide insights into how all of these processes interact and result in different patterns of deforestation across the Selva Lacandona.

Timber extraction has been widely implicated as a major cause of deforestation. From an historical perspective, the Selva Lacandona has been a rich source of mahogany and tropical cedar. In fact, during the period between 1864 and 1914, the Selva Lacandona was transformed into one of the most important economic regions in southeastern Mexico. Many have argued that the early penetration of foreign capital in the region initiated the deforestation process, which was then exacerbated when modern logging got underway in the 1960s. State-sponsored logging is held responsible for the clearing of large tracts of forest during the 1970s and 1980s. More recently, the process has been tied to clandestine logging by local farmers.

Oil exploration by Mexico's state oil company, PEMEX, has also been charged with incurring significant ecological damage, including deforestation. The presence of PEMEX camps in several parts of the forest led to concerns that the Mexican government had plans to develop the forest into an important oil-producing region, similar to the situation in Tabasco and Veracruz. PEMEX activities have been closely monitored, and actions ranging from the construction of roads to the seismic activities related to exploration have been criticized as damaging to the environment.

While timber and oil have been viewed as the objects of plunder in the Selva Lacandona, colonization of the forest by small-scale agriculturalists has been considered the most obvious example of irrational land use. Colonization has, in fact, had a very visible impact on the forest. The population of the Selva Lacandona region has increased dramatically over the past 30 years, from approximately 60,000 in 1960 to an estimated 350,000 in 1995. Many of the colonists migrated from other regions of Chiapas, particularly from Tzeltal, Tzotzil, Tojolabal, and Chol communities. Others came from impoverished Mexican states such as Oaxaca, Guerrero, and Puebla. Finally, some settlers came from private farms established on the periphery of the forest. Although most of the colonizers are members of *ejidos*,[1] some were able to purchase their own land to form small farms or ranches. A large proportion of migrants to the Selva Lacandona are members of indigenous groups.

Just as colonization is widely considered to be the direct cause of deforestation, population growth is often perceived as the principal force behind it. Population growth in the Highlands of Chiapas in particular is believed to have created land shortages that have pushed many members of indigenous communities into the Selva

Lacandona in search of land. High birth rates among women living in the Selva Lacandona are also considered a direct cause of deforestation.

The transformation of the landscape from forest to pasture leads many to point to cattle ranching as a leading cause of deforestation. The fact that cattle population of Ocosingo alone exceeded 100,000 heads provides strong evidence of the importance of pastures in this region. Both private ranchers and small-scale farmers have been implicated in what is often considered to be the most destructive use of tropical forest lands. The force behind cattle expansion is frequently attributed to the "hamburger culture" and a growing national and international demand for beef (Myers 1981; Nations and Komer 1987).

In short, the proximate causes of deforestation have led many to contend that population growth, poverty, and the introduction of non-native agricultural systems and cattle ranching to the fragile tropical forest ecosystem account for much of the deforestation, along with logging and oil exploration. Foreign-owned logging companies and the Mexican state have been directly implicated in the latter two activities. The appropriate responses related to these causes range from population control to increased economic opportunities in the poorer states of Mexico and other parts of Chiapas. Agricultural extension programs to "teach" colonists how to farm the forest have also been proposed, as well as ending state-supported contributions to deforestation through logging and curbing international demand for beef.

Each of these purported causes holds some degree of legitimacy, and can therefore be considered a component that helps to explain deforestation. However, a review of the direct causes falls short of providing a comprehensive understanding of the processes that have contributed to deforestation. Deforestation continues, despite changes in agricultural and extractive policies, such as the termination of credits for cattle production and a ban on logging in the forests of Chiapas. In fact, many of the perceived "solutions" to deforestation that respond uniquely to the direct causes may be actually frustrating efforts to preserve the region's remaining tropical forests. In the next section, the direct causes will be reexamined in the context of social, economic and political relations.

The Driving Forces

The proximate causes described above form a starting point for analyzing the driving forces of deforestation in the Selva Lacandona. Posing the questions *why? how?* and *where?* to each of the direct causes can lead to a deeper understanding of the social, economic, and institutional relations affecting deforestation patterns. However, the driving forces of deforestation cannot simply be listed or elaborated by attributing one or two underlying forces to each direct cause. Instead, the search for driving forces exposes a deeper and more complex set of roots that penetrate into the social and economic fabric of local, regional, national and global systems.

Each of the proximate causes takes place within an historical context. To understand *why* deforestation has occurred in the Selva Lacandona, one must examine the history and evolution of production activities, both within and outside of the region. In tracing the history of these activities, it becomes clear that deforestation is never an irrational activity: it always makes sense to some person or group. In fact, until quite recently deforestation was considered a sign of progress and development, rather than an ecological catastrophe (MacIntyre 1977).

To identify *how* deforestation results from production activities, one must go beyond a descriptive history of logging, colonization, or cattle ranching and analyze how access to land, labor and capital has changed as a result of policies or trends taking place at the local, regional, national and international scales. The relationships between these factors of production reveal a great deal about the impacts of various activities on the environment, and how they have changed over time.

Deforestation in the Selva Lacandona can also be explained by identifying *where* various production activities have taken place. Terrain, soils, roads, rivers, infrastructure, and other factors have influenced the location of colonization and economic activities, hence the patterns of deforestation. From a geographic perspective, the Selva Lacandona is not a homogenous entity, but a mosaic of subregions that maintain distinct characteristics. Focusing broadly on deforestation in the larger Selva Lacandona region, it is easy to overlook the locational context of deforestation.

Trees for Timber

The history of timber extraction in the Selva Lacandona is intriguing (see González Pacheco 1983 and de Vos 1988a), and it provides an example of how factors of production have varied over time and altered both the environmental and social consequences of logging. The importance of access to land, labor, and capital has been transformed as production systems have changed from selective logging by private companies to extensive logging by private companies and later by the state, and finally to selective logging carried out either legally or illegally by campesinos. Political relations have also formed an important part of this history, both in the relationships between capitalist entrepreneurs and national politicians and between small-scale farmers and local bosses representing the ruling political party. As a result of these factors, the impacts of logging on deforestation of the Selva Lacandona have varied both temporally and spatially.

Early logging in the Selva Lacandona was labor intensive, and little physical capital was used in the extraction of thousands of mahogany and tropical cedar trees from the region for export abroad. In contrast, financial capital and political connections were essential to enter into this industry. To profit from the region's tropical hardwoods, it was necessary to have access to the land through yield contracts, logging rights, or direct ownership (de Vos 1988a). Liberal land acquisition laws in Mexico during the second half of the 19th century made the national lands available to seemingly anyone interested in grabbing them. However, in reality the stipulations tied to claiming or surveying land made it extremely difficult for small-holders to accumulate more land, while it became increasingly easy for large landholders with political connections to expand their holdings.

Labor formed the backbone of the early logging industry, and the exploitation of a cheap labor force was essential to the profitability of the industry. The existence of such a labor force was not a mere historical anomaly. Instead, it was tied to a series of mechanisms developed by landowners and politicians in colonial and post-colonial Mexico (Benjamin 1989). Indians from the Highlands of Chiapas were particularly vulnerable to exploitation. Voluntarily or involuntarily, by the turn of the century they made up a large part of the labor force in the logging camps of the Selva Lacandona. Often tied to the logging companies through a system of debt-peonage, these workers were essential to the selective logging of millions of cubic meters of tropical hardwoods from the Selva Lacandona (Benjamin 1981).

The ecological impacts of early logging were relatively benign, as the logging was selective and permanent roads were not cut through the forest (de Vos 1988b). Most of the logged areas regenerated, and decades later there were few visible impacts of these earlier activities. More persevering, however, were the larger social structures that had facilitated this logging in the first place. While the owners of logging companies often sold off their land or moved their activities else-where, the labor force was left with neither land nor wage labor.

Although a few remained in the forest as pioneering colonists, most returned to their homes in search of other means to make a living.

Modern attempts at logging the Selva Lacandona began in the late 1950s, after several North American-financed companies attempted to initiate production in an extensive part of the region (González Pacheco 1983). The companies introduced mechanized logging to the region, and logged trees that were accessible only by cutting roads through the forest with heavy machinery. Access to land and capital were thus essential to the profitable extraction of hardwoods. In contrast, labor was only a minor consideration. These new endeavors opened up new areas to exploitation, and in the process incurred substantial collateral damage to the surrounding environment.

Due to the high expenditures associated with cutting roads through the forest to extract increasingly scarce mahogany and tropical cedar, private logging companies soon found that costs outweighed the profits that could be earned in the Selva Lacandona. In 1972, logging rights were sold to the state-owned Nacional Financiera, S.A. (NAFINSA), and in 1974 a parastatal logging company, *Compañia Forestal de la Lacandona, S.A.* (COFOLASA) was established. Access to timber resources was virtually guaranteed, as the government had titled 614,321 hectares of the forest to 66 Lacandón heads of family in 1972 (Diario Oficial 1972). The creation of the new Comunidad Lacandona was widely seen as a political maneuver to control timber resources that were being destroyed by the growing number of small-scale farmers colonizing the region (Nations 1979). Yet from an economic perspective, the state-run company was unsuccessful in exploiting the Selva's forest resources. COFOLASA proved unable to extract timber in a manner that was profitable or sustainable, and in 1989 terminated operations in the Selva Lacandona (Cruz Coutiño and Parra Chávez 1994).

The poor performance of COFOLASA reflected the general malaise of Mexico's forestry sector during the 1980s. In 1989, Chiapas governor Patrocinio González declared a state-wide ban on logging. The ban, known as the *veda forestal*, was declared as a concession to environmentalists struggling to slow deforestation in the Selva Lacandona. However, it could be considered a rather painless concession from the government's point of view, as logging no longer made significant contributions to the state's economy. The logging ban had its greatest repercussions on the peasants who extracted timber from their own lands, and it led to a number of conflicts in subsequent years (Fernández Ortiz et al. 1994).

With the exception of land belonging to the Comunidad Lacandona or protected as biosphere reserves and national parks, almost all of the region's remaining forests are associated with ejidal lands. Nevertheless, to date, few ejidos have been able to profit directly from the timber resources on their land, either because of a lack of the necessary equipment and marketing mechanisms, a lack of experience in the forestry sector, or other obstacles. Only a select number of ejidatarios have received logging concessions through political connections and have been able to harvest significant areas of the remaining forest (Harvey 1997). Legal and illegal logging operations have been particularly significant in the Marqués de Comillas region, raising the ire of influential conservation groups such as the *Grupo de los Cien.*

Although private logging companies are excluded from direct control of forest resources in much of the Selva Lacandona, changes to Article 27 of the Mexican Constitution made in 1992 allow ejidatarios to claim full property rights to collective forest resources, and to form legal associations with private enterprises to manage forests. Furthermore, private companies can buy and manage up to 20,000 hectares of forest lands (Segura 1996). More recently, a new forestry law was enacted in Mexico that provides some protection to ejidal forests by banning property transfers and encouraging efforts to develop community forestry (INE 1997). These changes may have implications for the development of the forestry sector in Chiapas, including a changing role for forestry on ejidal lands in the Selva Lacandona. Whether this law succeeds in revitalizing Mexico's crippled forestry sector remains to be seen.

Underlying Resources

Petroleum exploration provides another example of an extractive activity that has been widely implicated as a cause of deforestation. Oil reserves in the Selva Lacandona were discovered in the 1970s, at a time when international prices were extremely high. The potential reserves of the Selva Lacandona were viewed as a means not only for enriching the national treasury, but also as a way to integrate a marginalized region into the wider economy. Early exploration appeared promising, and plans were made by PEMEX to extract oil from a number of wells located in different parts of the region.

The deforestation associated with these exploratory wells has been relatively minor. In fact, the direct effects of PEMEX on deforestation have been limited to parts of the Marqués de Comillas subregion and around other drilling sites in the municipios of Ocosingo and Altamirano. In the Marqués de Comillas region, only about 130 hectares were cleared for roads and wells at seven drilling sites (Carmona Lara 1988). From an ecological perspective, the impacts have been more far-reaching, including water pollution, noise pollution, and an increased trade in endangered species.

The actual extraction of oil from the Selva Lacandona region was less than many anticipated, and for reasons that remain largely speculative, PEMEX terminated its operations in the region in 1994. Although the reserves of the Selva Lacandona are still thought to be considerably high, future exploitation depends more on Mexico's economic performance and the political situation in the region than on the anticipated ecological impacts of production activities.

Timber extraction and oil exploration have clearly contributed to deforestation in the Selva Lacandona. However, implicating extractive industries as directly responsible for deforestation over-simplifies the process to its physical manifestations, ignoring the complex web of social, economic, and institutional relations that have ameliorated or exacerbated their effects. Although the direct impacts of extractive activities are by no means trivial, the real legacy of logging and petroleum exploration has been the roads carved into the forest, particularly in the northern region near Palenque and Chancala, and from Palenque to Benemerito in the Marqués de Comillas region of the forest. These roads created apertures that facilitated colonization in what is considered by many to be one of Mexico's last agricultural frontiers.

The roads that have been constructed to gain access to extractive resources in remote areas have indeed served as conduits for colonization in some parts of the Selva Lacandona. Nevertheless, it would be a fallacy to ascribe colonization to the presence of roads alone. Many of the roads were actually built along existing networks of footpaths and trails that connected inhabitants to regional markets and the "outside world." In areas that were not significantly affected by logging, such as along parts of the Rio Santo Domingo in Las Margaritas, the roads that exist today can be more accurately considered a result of colonization, rather than a cause.

From Forests to Fields

Colonization of the Selva Lacandona by small-scale agriculturalists reflects much more than a problem of population growth and an associated lack of agricultural land in other parts of Chiapas and Mexico. To understand why the forest has been colonized, one must look

at agrarian relations in Chiapas in the context of agricultural transformations in Mexico. The current century has witnessed dramatic
changes in Mexico's agricultural sector, accompanied by some rather
conspicuous continuities within Chiapas. While the transformations
have led to a present-day crisis among rural producers throughout
Mexico, the situation in Chiapas has been exacerbated by outdated
agrarian structures and the development of agriculture along two
divergent trajectories. These trajectories reflect the legacy of historical
tensions between traditional landowners of the Highlands and
entrepreneurial landowners of the Central Valley. The agricultural
system that emerged from this split resulted in the juxtaposition of
highly productive export enclaves that have been important factors in
Mexico's agricultural productivity, alongside underproductive agricultural lands turned over increasingly to cattle ranching (Benjamin 1989;
Reyes Ramos 1992).

The roots of agrarian relations in Chiapas can be traced to the
colonial period, when the labor of the largely-indigenous population
was appropriated to serve the accumulation of wealth among Spanish
colonial administrators and clergy. During this period, some of the best
lands of indigenous communities were incorporated into the colonial
system (Marion Singer 1988). Upon unification with Mexico in 1824, the
colonial ruling class of Chiapas was replaced by a local elite who
continued to reinforce the existing structures. A number of laws
facilitated an extensive transfer of indigenous lands into the hands of
non-Indian elites. Displaced Indians were then incorporated into a
labor force that was created and maintained through a variety of
systems, including sharecropping, tenant farming, wage labor, and debt
peonage (Wasserstrom 1983).

During the Porfiriato period, the number of large landholdings grew
at an astounding rate, and more and more of the state's agricultural
workers found themselves in the position of day wage laborers,
sharecroppers, or renters (García de León 1984b; Benjamin 1989). Many
of these workers were tied to landowners through either debt or
paternalistic relationships, hence migration to other areas in search of
land was rarely considered a viable option. The Mexican Revolution
that liberated many peasants from large estates in other parts of
Mexico took on different characteristics in Chiapas. In fact, most
peasants in Chiapas joined forces with the landowning class to fight
nationalist policies aimed at breaking up large estates and ending labor
exploitation. In a sense, they were defending what had until then been
their only option for sustenance (García de León 1984b; Benjamin 1989;
Reyes Ramos 1992). The Mexican Revolution thus bypassed Chiapas,

and set the stage for atavistic agrarian policies that continue to shape the situation in the state today.

The counterrevolutionaries triumphed in Chiapas after the Revolution, and firmly entrenched the landholding elite's position in state government. Laws were passed that established the maximum size of property at 8,000 hectares, one of the highest ceilings in Mexico (García de León 1984b; Reyes Ramos 1992). Furthermore, plantation agriculture was excluded from agrarian reforms by federal legislation, exempting a large part of the state's most productive lands. Post-revolutionary reforms were thus extremely weak in Chiapas. It was not until 1934, when the Cárdenas administration made agrarian legislation the exclusive domain of the federal government, that the first significant reforms took place. These reforms happened to coincide with a period of growth in union activities and socialist organizations (Benjamin 1989). Despite the minimal extent of the reforms, landowners viewed them as a direct threat, and organized among themselves to counter any changes that might erode their privileges. Land tenure still remained highly concentrated in 1940, and the struggle between landowners and peasants had begun (Reyes Ramos 1992).

During the 1940s, when agricultural production was taking off in much of Mexico, the situation in Chiapas was remarkably stagnant. The export enclaves that produced commercial crops such as coffee, cacao, and bananas were thriving, whereas the lands held by the conservative elite were underutilized. Over one million hectares of land were in the hands of administrators rather than landowners, who often viewed their holdings as sources of perpetual rent rather than as engines of agricultural growth (Reyes Ramos 1992). Forced to address problems in the agricultural sector, the government eventually turned to land distributions. Many of the earliest distributions were located near coffee plantations to ensure a supply of labor, or near the border with Guatemala to establish a visible presence (Reyes Ramos 1992). Most of the later distributions involved national lands, hence the forests of the Selva Lacandona region became an indirect target of the state's agrarian strategy.

Chiapas possessed about 3 million hectares of national lands in 1940, which under Mexican law could be colonized by individuals, as ejidos, or as communal lands (Reyes Ramos 1992). These lands provided the state government a wide degree of latitude with which it could carry out the impending distributions. Rather than disturb the interests of existing landholders, the government was able to address agricultural stagnation and satisfy the growing demand for land by opening up new areas for colonization. The earliest benefactors of these distributions of national lands were individuals, but by the early 1960s land distributions were directed toward the creation of *Nuevos Centros de Poblacion Ejidal* (NCPEs) (Paz Salinas 1989).

The social structures that have historically enabled some to accumulate large estates while alienating others from the land and appropriating their labor has led to a political impasse that was earlier resolved by opening up new areas to cultivation. Agrarian structures were not significantly affected by these distributions, and the landowning class remained intact (Benjamin 1989; Reyes Ramos 1992). Yet while agrarian politics in Chiapas have changed little over the past century, Mexican agriculture has undergone dramatic transformations (Grindle 1986; Sanderson 1986; Barkin 1990; Barry 1995). Agriculture has become increasingly commercialized and capitalized, with large and modern farms responsible for much of the sector's productivity. The so-called social sector of agriculture has become increasingly irrelevant from both a production and policy perspective.

Agricultural change in Mexico has taken the form of an increasing commitment to export production, based on costly inputs and cheap labor. The majority of farmers in Mexico, however, work small plots of land that are often of marginal quality. Populist policies of the 1970s were aimed at increasing production of basic grains on ejidos and small landholdings, but such policies have long disappeared as a result of the continuing financial crises that have beset Mexico since 1982 (Hewitt de Alcántara 1994). More recent agricultural policies have eliminated price controls, subsidies, and other state-sponsored programs directed at small-scale producers. Along with Mexico's neoliberal trade policies that will eventually open national markets to cheap food imports from the north, these changes deprive small-scale producers of the social and economic safety nets that were becoming increasingly essential for survival (Barry 1995; Harvey 1995).

It is within the context of agricultural transformations in Mexico that colonization of the Selva Lacandona over the past thirty years must be considered. Much of the land in the Selva Lacandona region was nationalized during the 1940s and 1950s, offering a convenient outlet for channeling land distributions. Although agrarian authorities directed peasants toward the Selva Lacandona, colonization has been for the most part spontaneous, with settlement followed by a long series of bureaucratic maneuverings to gain official title to the land (Paz Salinas 1989; Reyes Ramos 1992). The government rarely helped to develop infrastructure to facilitate colonization or subsequent agricultural production. Instead, most colonists walked for days to reach their communities, or paid for costly air passage in small planes that could land at the airstrips associated with the numerous climate stations operating in the region. Logging roads expedited the process in some parts of the forest, while roads constructed by PEMEX facilitated colonization in other parts.

By 1990, there were 568 ejidos in the Selva Lacandona region, as well as almost 2,500 *localidades*, or settlements of various sizes (Leyva Solano and Ascencio Franco 1996). The patterns of these settlements reflect a very distinct history of colonization and help to explain the system of social and economic relations that emerged in the region, including the increasing social stratification among and within communities (Lobato 1980; Leyva Solano and Ascencio Franco 1996). Although the earliest arrivals suffered great hardships in colonizing the humid, tropical environment, they also held the option of selecting these productive lands (Garza Caligaris and Paz Salinas 1986; Paz Salinas 1989; Leyva Solano and Ascencio Franco 1996). The most productive lands were located in fertile valleys close to sources of water. The earliest settlers were also able to appropriate the labor of more recent arrivals, and thus they could more readily expand their production.

The land along rivers closest to the municipal centers of Ocosingo and Las Margaritas were among the earliest areas to be settled, with colonization taking place during the 1950s and 1960s. The northern portion of the Selva Lacandona region was also colonized during this period. However, as colonization progressed, more remote areas were cleared for settlement, followed later by more marginal areas on slopes, mountain tops, or in locations distant from water supplies. One of the most recent areas to be colonized was the Marqués de Comillas subregion, including the upper reaches of the Rio Lacantún, just below the narrow Cañon del Colorado. These areas, which were among the most difficult to access from Chiapas, formed part of the sparsely inhabited frontier area along Mexico's border with Guatemala. It was not until the late 1970s and 1980s, when these lands were colonized, that deforestation became visible.

Land distributions in the Selva Lacandona were not part of a government strategy to increase agricultural production. Instead, it served as a means to satisfy demands for land and maintain the large estates elsewhere in Chiapas that were considered essential to commercial agricultural productivity. Land distributions also contributed to a greater presence along Mexico's southern border, which was considered to be important when Guatemalan civil war erupted.

Cultivating the Forest

Given the political nature underlying colonization, it is not surprising that migrants into the Selva Lacandona are often accused of mismanaging the land, importing production systems that are unsuitable in the hot and humid tropical environment. Many compare contemporary agricultural practices in the Selva Lacandona with the traditional agricultural systems of the Lacandón Indians, which have proven to be excellent examples of sustainable use of tropical forests (Nations 1979; Nations and Nigh 1980). Although the production systems of the Lacandones are unique and merit special attention, those of recent immigrants have actually evolved quite well to meet the demands of local conditions (Price and Hall 1983). In fact, the farming systems can be considered dynamic and capitalistic, despite a reliance on rudimentary tools and human labor (Lobato 1980).

Farming practices vary within communities and regions, but generally include a combination of subsistence agriculture, semi-commercial agriculture, and commercial agriculture (Price and Hall 1983). Slash-and-burn cultivation dominates, with a cropping mix of maize, beans, squash, bananas and plantains, coffee, chile, cacao and sesame, alongside livestock production. Although the primary objective is to maintain subsistence, most farmers in the Selva Lacandona strive to produce a marketable surplus. However, there are a number of constraints that hinder productive agriculture, including a scarcity of production inputs and underdeveloped marketing mechanisms. The failure of agricultural systems to provide a secure income for farmers in the Selva Lacandona has led many of them to turn to cattle production for cash income (Price and Hall 1983).

Cattle ranching as a production activity took off in Chiapas after 1950, replacing both agricultural and plantation crops (Fernández Ortiz et al. 1994). Between 1940 and 1980, a series of policies and programs initiated by the state and national governments provided supports and incentives to cattle ranchers, and protected cattle estates from agrarian reforms. As agricultural lands throughout Chiapas were converted to pasture, the campesinos working the land lost their jobs and joined the growing unemployed rural population. In contrast to agriculture, cattle ranching is relatively inexpensive in terms of both capital and labor, but costly in terms of land. With the exception of a few large ranches established within the Selva Lacandona during the 1940s and 1950s, most of the cattle occupied formerly agricultural lands throughout Chiapas. However, after 1970 cattle production began to expand into the forested lands of the Selva Lacandona. Landowners would allow campesinos to clear the forest and cultivate the land for several years. After that period, grasses would be planted and agriculture would eventually be squeezed out of the production cycle (Leyva Solano and Ascensio Franco 1993).

At the same time that private ranchers were expanding into the Selva Lacandona region, ejidatarios and small landholders also began transforming forests and formerly forested areas to pasture lands. Cattle provide farmers with a higher income in a shorter period of time, allowing them to produce beyond levels of subsistence and acquire cash income (Price and Hall 1983). Cattle production also holds benefits for the community, providing income for public works projects or money to cover travel to the state capital to take care of agrarian issues (Leyva Solano and Ascensio Franco 1993). Cattle expansion has been most rapid in areas of the Selva Lacandona where credit has been readily available, and least rapid where credit has been absent (Price and Hall 1983). Credit has generally been more readily available to some of the more established ejidatarios who had developed a capital surplus in agricultural production. The mere anticipation of forthcoming credit has also lead to the establishment of pastures, which have remained regardless of the arrival of credits. After reaching a peak in the mid-1980s, credits for cattle ranching became increasingly scarce. Although the cattle industry of Chiapas has contracted dramatically over the past decade, cattle production remains an important activity in the Selva Lacandona (Villafuerte Solís et al. 1993: Fernández Ortiz et al. 1994).

While cattle have become an increasingly common sight in the Selva Lacandona region, the impacts of cattle ranching on the environment are variable. Although each head of cattle requires approximately two hectares of land, some pastures are considerably overstocked. Small-scale cattle operations have a relatively minor impact on the region's ecology, since typically the least productive lands are planted with grasses (Price and Hall 1983). However, large-scale ranching as practiced by smallholders or large landholders generally impacts extensive areas, and is sustainable for only a few years. The length of time that cattle grazing is sustainable is dependent on pasture management. If land is grazed for too long, it becomes increasingly difficult for tropical forests to regenerate.

Political Conflicts

Timber extraction, oil exploration, agricultural expansion, and cattle production are examples of the ways in which economic, social, and political relations at various levels of analysis have influenced the destruction of tropical forest. Each factor has had significance for distinct subregions in the Selva Lacandona during different periods of time. These same relations have also given rise to political conflicts that have had profound impacts on parts of the forest. The earliest of these was the guerrilla war in Guatemala that resulted in thousands of peasants seeking refuge in the Selva Lacandona during the 1980s (Manz 1988). The second political upheaval was the Zapatista Uprising that began in 1994 and has brought international attention to the region, as well as an expanded military presence (see Collier and Quaratiello 1994; Katzenberger 1995; Ross 1995; Conpaz et al. 1996).

The underlying causes of the guerrilla war in Guatemala are tied to the unequal land distribution, labor struggles, and treatment of the country's largely indigenous population. As nonviolent grassroots organizations were increasingly repressed by the government in the 1970s, radical guerrilla movements including the *Organizacion Revolucionario de Pueblos en Armas* (ORPA) and the *Ejército Guerrillero de los Pobres* (EGP) formed as an alternative means of addressing the country's economic and social injustices. The response of the Guatemalan army was to eradicate guerrillas by killing all potential supporters. The brutal tactics used by the army to intimidate and destroy the communities that had settled in the forest led to a massive flow of refugees across the border into Mexico (Manz 1988; Falla 1994). By 1984, over 46,000 refugees were settled in 90 camps in Chiapas, with some of the largest camps located in the Selva Lacandona, close to the border with Guatemala (Manz 1988).

The presence of refugees in the Selva Lacandona had some direct impacts on the forest in the areas along the Mexico-Guatemala border. For example, land was cleared to establish camps and cultivate crops to supply food for the refugees. More significant was the impact of a cheap labor supply on local economic relations: refugee labor was appropriated to clear land, prepare fields, and cultivate labor-intensive crops such as coffee. As an outcome of the war, the Mexican government also constructed a highway along the border, and encouraged Mexicans to colonize the area. Between 1980 and 1986, 10 ejidos were established along the border in the Marqués de Comillas region to establish a stronger Mexican presence (González Ponciano

1990). As a result of these factors, deforestation increased dramatically.

The more recent Zapatista Uprising, which began in 1994, but whose roots in the area can be traced back to the early 1980s, has also played an important role in reconfiguring the Selva Lacandona. Many of the underlying causes of the uprising are related to the social relations that have driven extractive industries and colonization in the Selva Lacandona. Yet the Zapatista struggle for dramatic social, economic and political changes in Mexico has inadvertently led to greater deforestation, particularly through extensive militarization of the Selva Lacandona and accelerated road construction. The politics of appeasement has enabled some campesinos to resume logging in areas that are considered "neutral," and others have taken advantage of the tense political situation and invaded land in protected ecological areas (Nations 1994). Although the Zapatistas are concentrated largely in the western part of the forest, indirect impacts of the uprising extend to the eastern and southern portions of the Selva Lacandona, particularly along the Rio Usumacinta and Rio Lacantun.

Saving the Selva

The role of conservation efforts in configuring the present day forest is critical to understanding the political ecology of deforestation. The core of the remaining forest lies in the 331,200 hectares of the Montes Azules Integrated Biosphere Reserve, which was officially decreed in 1978 (Diario Oficial 1978). In 1992, Montes Azules was supplemented by several other protected areas, adding 74,000 hectares to the total area of protected land in the region (Diario Oficial 1992a; Diario Oficial 1992b; Diario Oficial 1992c; Diario Oficial 1992d).

The establishment of these reserves can be credited to scientists and environmentalists dedicated to saving one of Mexico's most valuable tropical forests (Vásquez Sánchez 1992). International environmental organizations such as Conservation International have also played an important role in securing resources for conservation, and the World Bank's Global Environmental Facility has targeted Montes Azules as an one of Mexico's most important protected areas in need of management (GEF 1992). Nevertheless, the future of the region's biosphere reserves and protected areas is in a precarious state, largely because there are few linkages between social and environmental struggles in the Selva Lacandona region.

The land under protection in the Selva Lacandona has been the focus of an increasing number of conflicts, in part due to the fact that some communities had settled in the area prior to the declaration of the biosphere reserve, and in part because some perceive the reserve as a means of enclosing much needed land and ignoring local needs. Almost 10,000 people were living within the Montes Azules reserve by the early 1990s, and as many as 13,000 were living directly outside of its limits (March 1994). These numbers have increased dramatically over the past several years, partly as a consequence of the Zapatista Uprising.

The Mexican government has encouraged rather than alleviated tensions between local people and environmentalists by pursuing contradictory policies related to tenure within the reserve. Although some institutions within the government are committed to conservation, others are solely concerned with resolving tenure issues. The situation reflects a lack of policy coordination among government institutions. More cynically, one could argue that both sides are aware of the contradictions, but unwilling to address the complexities involved in resolving the matter.

Although efforts to preserve the remaining tropical forests of the Selva Lacandona are often viewed with suspicion or considered by locals to be an elitist project, the protected areas have to a large extent determined the present-day configuration of the forest. Where the reserves are bounded by rivers, there is an abrupt change in land cover on the reserve side. Where the borders are less well-defined, the colonization process encroaches on the reserve, eating away at Mexico's largest remaining tropical forest.

Almost all of the available land in the Selva Lacandona has already been titled or claimed. Population growth within the Selva Lacandona is above the state average, putting pressure on both ejidal forest reserves and nationally protected areas. Many consider biosphere reserves as the last remaining frontier for colonization. Although the outcome of government negotiations with the Zapatistas has yet to be determined, there is little guarantee that Montes Azules and other protected areas in the Selva Lacandona will not be compromised to the growing demand for land and the increasing integration of the region into the state, national, and global economies. Tensions between environmental and social objectives are thus mounting in the Selva Lacandona region.

Weaving it Together

The brief introduction to the political ecology of the Selva Lacandona region presented in this chapter brings out some of the complexities in the driving forces of deforestation that are absent from discussions that focus only on the proximate causes. The growing recognition that these complex driving forces are key to understanding the process of tropical deforestation is symptomatic of a paradigm shift in deforestation studies. These forces will be elaborated on in the following chapters, with an emphasis on the heterogeneity of the Selva Lacandona region.

A political ecology approach views deforestation as a complex and multifaceted issue closely tied to social and economic relations operating across scales, from the local to the global. In fact, "population, human welfare, and environmental degradation are dimensions of the same complex problem" (Arizpe et al. 1994:6). From this perspective, the causes of deforestation in the Selva Lacandona penetrate much deeper than simple population growth. As Arizpe et al. (1994:3) emphasize;

> 'the population problem' does not just involve absolute numbers of people nor even just population densities or overall rates of increase, but also, in important ways, social, political and institutional factors. Complex patterns of human relationships overlay, alter and distort the relation of people to the land and to the cities.

Weaving together the political ecology of the region will provide important insights into the growing tensions between environmental and social struggles in the Selva Lacandona, particularly those occurring at the interface between protected areas and surrounding communities. Conservation efforts can be credited with maintaining a large part of the remaining forest and expanding awareness about deforestation. Likewise, social struggles have brought international attention to the marginalized communities of the Selva Lacandona, as well as visible changes such as electricity, solar energy, and improved roads. Nonetheless, the distinction between environmental struggles and social struggles is emerging as one of the most critical challenges facing Chiapas today. The remaining chapters will describe how these struggles developed along distinct lines, and suggest how they might be resolved.

3

Reconfiguring the
Selva Lacandona

A Region Transformed

The Selva Lacandona was once a vast expanse of tropical forest, dissected by numerous streams and rivers that feed into the largest river in Mexico, the Río Usumacinta. Early maps from the 19th century show this region to be a large, unexplored area referred to as the "Desert of Solitude." If satellites had existed in the mid-1800s, a resulting image would have depicted a dense green forest, with few detectable signs of human settlements. One would have noted the mountainous terrain and the intricate network of rivers, and a keen eye might have differentiated the forest types and observed how they vary with elevation and proximity to water. The Selva Lacandona of 1850 was familiar to the small groups of Lacandón Indians living within, and formidable to those living outside of it.

The Selva Lacandona of the 1990s invokes quite a different picture. The region is home to over 300,000 people, along with almost an equal number of cattle (INEGI 1993). Roads encircle the Selva Lacandona and penetrate all but the core of the remaining forest, and a flow of buses and pick-up trucks connects the Selva with commercial centers in Chiapas, Tabasco, and Guatemala. Electric wires have been dragged across expansive areas to provide electricity to many of the small communities, and satellite dishes and televisions are becoming increasingly visible in remote areas. Along many of the valleys and riverbanks of the Selva Lacandona, the forest has been replaced by agricultural fields and cattle pastures. The satellite images that capture the land cover of the 1990s show a frontier agricultural region with forests increasingly restricted to the crests and slopes of mountains, along with a core of the forest within the protected confines of the Montes Azules Biosphere Reserve.

Over the past century and a half, the Selva Lacandona has been dramatically reconfigured. The name "Selva Lacandona" no longer describes a vast forest cover, but instead refers to a region that is undergoing rapid changes. The loss of tropical forest cover describes only one aspect of the changes taking place, but because of its significance for the ecology and biology of the region, it is considered a critical one. To interpret the complexity of relations contributing to the deforestation process, it is essential to understand the region from an historical context. It is also important to gain a perspective on the rates and patterns of land cover changes, as these are closely related to the social causes of deforestation. In this chapter, the recent history of the Selva Lacandona will be reviewed, and various estimates of deforestation will be considered. An original estimate of deforestation will also be presented, based on an interpretation of satellite imagery. Variations in the patterns of deforestation in five subregions of the forest will be examined, in order to illustrate the heterogeneous nature of deforestation.

The Historical Context

The Selva Lacandona has a rich Maya history that has transformed the area into a prominent center for archaeological and anthropological studies.[1] Over 200 ruins attest to the importance of the Selva Lacandona as a population center during the Classic Maya period, which ended about 900 A.D. During this period, extensive areas of the forest were cleared to support a variety of agricultural systems, including slash-and-burn rotational cultivation (Casco Montoya 1990). These systems supported population densities ranging from 4 to over 300 persons per square kilometer (Casco Montoya 1990). One theory holds that deforestation contributed to the collapse of the Maya Empire (Abrams et al. 1996). The tropical forest eventually regenerated and buried evidence of earlier civilization. Forests so successfully reclaimed the area that many people erroneously perceive the current Selva Lacandona as virgin rain forest.

At the end of the Classic Maya period, a number of groups were scattered throughout the Selva Lacandona, including the western Chol and Cholti Indians. Collectively, these two groups were called Lacandones by the Spaniards (Nations 1979). Upon arrival in Chiapas, the Spanish introduced new diseases that decimated the Indian populations. Resettlement efforts were then undertaken in an effort to exploit the remaining indigenous populations for their labor, and benefit the colonial economy. A series of military incursions undertaken between 1559 and 1694 decimated the populations of Lacandones in the Selva Lacandona. Beginning in 1697, new groups of Indians from Campeche and the Petén migrated to the Selva Lacandona. These groups, ancestors of the present-day Lacandón Indians, spoke a different language than the earlier Lacandones (Nations 1979; de Vos 1980). The new Lacandones settled in dispersed communities in an area that was considered by most outsiders to be a remote and inaccessible jungle.

Deforestation in the modern Selva Lacandona is largely a recent phenomenon. In fact, it has been estimated that almost 90 percent of the total deforestation has taken place within the past three decades (Fuentes Aguilar and Soto Mora 1992). Nevertheless, to understand the contemporary cycle of deforestation, one must begin with the early explorations in search of ways to access tropical hardwoods for the international timber trade. This point of departure is one which "marks the incorporation of the riches of the Selva into the gears of capitalist accumulation" (González Pacheco 1983:51, own translation).

Although the Selva Lacandona has been closely identified with Chiapas during the twentieth century, its earlier history is in many ways more appropriately related to events taking place in Guatemala or in the neighboring state of Tabasco. During the 19th century, the Selva Lacandona was merely one part of a larger forest that extended into Tabasco, Campeche, and Guatemala. The state boundaries within this larger forest were poorly defined, and the international borders with Guatemala were for a long period under dispute (de Vos 1988a). For the *chiapanecos*, the Selva Lacandona was considered a desert — a vast region poorly represented even on the most up-to-date maps. As late as 1856, the Selva Lacandona appeared on official state maps as the *"Desierto incógnito habitado por los indios lacandones"*[2] (de Vos 1988a:16). This describes an area which was populated by several hundred Lacandón Indians, yet for the most part remained uncharted territory.

While official knowledge about the region was limited, locals in the areas around Palenque and Ocosingo were aware of the oral histories passed down from early colonial military and missionary ventures into the forest. The fertile valley of Ocosingo, where at least nine haciendas of the Dominican priests of Ciudad Real[3] were located, was long considered the gateway into the Selva Lacandona (de Vos 1988a). Nevertheless, the dense tropical forests and mountainous terrain inhibited most explorations, and as a result the Selva Lacandona remained beyond penetration from the western side. In contrast, the forest was much more accessible from the state of Tabasco, along the higher reaches of the Río Usumacinta near the town of Tenosique. This entryway had been used for years to facilitate trade with the Guatemalan department of Alta Verapaz. The river was not navigable in all parts, therefore *tenosiqueros* were forced to cut a path through the forest to bypass the dangerous sections. Upriver from the archaeological ruins at Yaxchilán, the Usumacinta River offered easy passage to and from the Petén of Guatemala (de Vos 1988a). It was, in fact, from the Petén region that the Selva Lacandona could be most readily reached.

Despite the traffic along the Usumacinta river, the interior of the Selva Lacandona remained distant and unexplored. It was regarded as the exclusive territory of the Lacandón Indians, who were feared as human-eating savages (de Vos 1988a). Nevertheless, a few outsiders did venture into the forest during the first half of the nineteenth century, with the objective of penetrating the Selva Lacandona. There were several motivations for these early excursions, primary among them being a search to discover a means of extracting tropical timber — a navigable river route in particular. Other reasons included the search for more direct trade routes to the ports of Tabasco, or sheer adventurism.

One of the earliest modern ventures into the Selva Lacandona took place in 1826, when Cayetano Ramón Robles of Ciudad Real led an expedition into the Selva in search of a passage for agricultural goods between Ocosingo and the ports of Tabasco. The expedition failed on two attempts, consequently the official report concluded that the Río Jataté was not navigable. Furthermore, the report suggested that the tropical hardwoods along the Río Jataté were neither abundant nor readily accessible. As a result of this negative account, the governments of Mexico and Chiapas lost the little interest that they had in the exploitation of timber from the Selva Lacandona by way of Ocosingo (de Vos 1988a).

In subsequent years, numerous other attempts were made to enter the Selva Lacandona from Chiapas, and each confirmed the earlier findings. In the 1850s, the governor of Chiapas granted a group of North Americans permission to undertake an expedition along the Río Santo Domingo. They failed to return, and nothing more was heard of them. In 1864, a military escort consisting of soldiers stationed in Ocosingo ventured down the Río Jataté, as did the political leader of Chilón in 1868 (de Vos 1988a). Neither of the expeditions succeeded, and as a result the Selva Lacandona remained out of reach from Chiapas.

Individuals with entrepreneurial spirits continued to believe that tropical woods could be profitably extracted from the Selva Lacandona, and two people in particular should be recognized in their attempts to "discover" the forest. The first, a *tabasqueño* named Felipe Marín, in 1859 suggested to the chief magistrate of the Petén in Guatemala that an experiment be made to test the possibility of transporting logs along the river from the upper Usumacinta to the state of Tabasco, passing through the rapids below Yaxchilán. The success of this experiment would open up vast new areas for timber exploitation. Marín asked for and was granted permission to cut 70 mahogany and tropical cedar trees. In 1860, 72 trees were cut from both sides of the Usumacinta near the confluence of the Río Lacantún. Months later, when the river rose with the first rains, 70 of the 72 trunks were retrieved south of Tenosique, rendering the experiment a great success (González Pacheco 1983; de Vos 1988a). As a result of this experiment, the logging companies of Tabasco gained access to a new source of tropical timber as far as the Lacantún river basin.

The western portion of the Selva Lacandona was explored by Juan Ballinas, the owner of the hacienda "El Paraíso" in the valley of Ocosingo. Ballinas was obsessed with the possibility of exploring the vast forest to the east of his hacienda, what he referred to as "*mi desierto*" (Ballinas 1951:18). Guided by the motto "*querer es poder*,"[4] he made several excursions into the forest between 1874 and 1875 with his friend and neighbor, Manuel José Martínez. Realizing that further exploration would require financial backing, he wrote to a timber baron in Tabasco in search of support, offering in return the rights to the vast supply of tropical woods of the area. The timber baron, Policarpo Valenzuela, took advantage of the offer and sent one of his sons on an expedition with Ballinas. The party that embarked on the adventure on March 7, 1877 did not get very far down the Río Jataté, but far enough to adequately convince Ballinas of the possibility of going further (de Vos 1988a).

Later that year, Ballinas led one more expedition into the Selva Lacandona, which this time turned out to be a success. Armed with a grandiose title bestowed upon him by the governor of Chiapas, "Discoverer of the Desert to the East of Ocosingo," in late 1877 Ballinas made it down the Jataté to the Río Lacantún, and from there to the Usumacinta and Río de la Pasión in the Petén region of Guatemala. Along his voyage, Ballinas was surprised to note that tabasqueño loggers were already established on the banks of the Río Lacantún. Even more to his surprise, he was arrested in the village of Sacluc for having invaded Guatemalan territory. At this time, all of the land in the present Marqués de Comillas region and to the south of the Río Lacantún belonged to Guatemala (de Vos 1988a).

It is important to emphasize that initial interest in the Selva Lacandona emerged largely from Tabasco, rather than from Chiapas. The persistent interest in the timber of the Selva Lacandona on behalf of businessmen from Tabasco can be explained by historical antecedents. Mahogany (*Swietenia macrophylla*) had become a highly sought after hardwood during the 17th and 18th centuries for use in ship construction, as well as for fashionable English furniture styles such as Chippendale and Sheraton (de Vos 1988a). The European mahogany markets were controlled through London and Liverpool, where the different varieties were recognized and classified according to their origin.

By the mid-19th century many of the original mahogany sources in the Caribbean, particularly Jamaica, had been exhausted. Recognizing the economic implications of this, English importers decided to open up new areas for exploitation. Tabasco provided the most promising opportunities, as it was a coastal state with an extensive river system. Thus, in addition to Belize, Honduras, and Nicaragua, the coast of Tabasco became an important supplier of mahogany for the European markets. However, it soon became obvious that the trunks from inland sources of Tabasco were of much better quality than the coastal specimens. In fact, the quality of the trunks rivaled the highest-grade specimens from the Caribbean. More important, the Selva Lacandona was the last intact reserve of mahogany in Mexico and Central America. By about 1870, English importing companies as well as timber firms in San Juan Bautista, Tabasco, decided that the time had come to open up the Selva Lacandona for timber exploitation (de Vos 1988a).

Liberal land acquisition laws passed during the second half of the nineteenth century placed the Selva Lacandona under the ownership of a handful of land speculators, foreign investors, and timber companies from Tabasco. As discussed in the following chapter, the politics of land ownership were much more complicated than simple land grabs. Instead, titles were conditionally granted until certain requirements were fulfilled. There were procedures to be followed, and often those with political power, money, or access to foreign capital were able to seize the opportunities. The national government could rescind titles at its discretion, and in many cases it did. However, landowners could petition against the decisions, and often political connections led to the restoration of titles. The situation became more critical in the post-revolutionary period, when many lands titles were annulled by the government for improper entitlements, tax evasion, or a number of other reasons.

Throughout the second half of the twentieth century, much of the land in the Selva Lacandona was returned to the national government or sold as large estates (González Pacheco 1983). The areas that passed directly into the hands of private landholders were located primarily in the northern municipio of Palenque, where cattle ranchers were amassing large tracts of forest to convert to pasture. Other lands were granted to individuals or groups of peasants by the Mexican government as part of the agrarian reforms. Finally, a large portion of the forest was legally titled to the Lacandón Indians by presidential decree in 1972. As described in the next chapter, this move enabled the government to secure timber concessions and continue the extractive industry that private, foreign-controlled capital had initiated one hundred years earlier.

This brief review of the early history of the Selva Lacandona introduces an important point that will be expanded upon in the following chapters. The early exploration and logging activities that took place within the Selva Lacandona did not have major impacts on the forest cover. In fact, not much deforestation occurred prior to 1960, with the exception of the northern fringes of the Selva Lacandona, around Palenque. However, the land and labor relations that characterized property acquisitions and the early logging industry were extremely important, in that they set the stage for a variety of social processes leading to deforestation. The patterns of deforestation that have emerged over the past four decades are not random patterns, but instead reflect the ways that production relations at local, state, national, and international scales have been played out over the past 130 years in the Selva Lacandona.

Estimates of Forest Loss

The Selva Lacandona of 1997 bears little resemblence to the Selva Lacandona of 1850 or 1960. Flying over the region, one can only imagine it as a vast, remote jungle perceived to be a desert of solitude. On the other hand, one views enough forest cover to question the assertions that three-quarters or two-thirds of the forest have been lost, and that the region is becoming a large desert, not of solitude, but of eroded pasture. Instead, the Selva Lacandona appears to be a region that is continuously being reconfigured, with large tracts of forest remaining in some areas, and little forest remaining in others.

The loss of tropical forest cover in the Selva Lacandona should be considered within the context of national deforestation statistics for Mexico. Estimates of deforestation in Mexico vary widely, ranging from between 370,000 hectares per year to 1.5 million hectares per year, depending on the definitions, data sources, and methods of calculations (Masera 1996). Synthesizing an assortment of studies and methodologies, Masera (1996) estimated that approximately 320,000 to 670,000 hectares of closed forest were destroyed each year between the end of the 1980s and beginning of the 1990s.

Of the 200,000 km^2 of tropical forests that existed in Mexico at the turn of the century, it has been estimated that less than 10 percent remains today (Pérez Gil 1991). Most of the tropical forests along the Pacific coast and Gulf of Mexico were deforested during the last century, as a result of logging, colonization, or industrial development. Remnants of these forests remain in Jalisco, Nayarit, Veracruz and Oaxaca. In addition, extensive areas of semi-deciduous or dry tropical forests can be found on the Yucatan Peninsula. The largest area of remaining tropical rain forest in Mexico is, however, found in the Selva Lacandona of Chiapas.

The Selva Lacandona remained relatively intact until the 1960s. Since then, however, the landscape has changed dramatically. It has been estimated that half of the original forest was destroyed between 1875 and 1988, with 585,000 hectares lost between 1969 and 1988, in contrast to a mere 80,000 hectares deforested over the previous 94 years (Pérez Gil 1991). In the 1969-1988 period, the average rate of deforestation amounted to 32,000 hectares per year. Another similar estimate, based on a study by Calleros and Brauer (1983) suggests that 584,178 hectares of land was destroyed by 1983. Based on an original forest cover of 1.3 million hectares, this amounts to a 45 percent reduction in size (Lazcano-Barrero et al. 1992).

A more extreme estimate, attributed to Jeffrey Wilkerson, indicates that 70 percent of the forest was lost between 1960 and 1991 (Fuentes Aguilar and Soto Mora 1992). Between 1875 and 1960, the 1.3 million hectares of forest was reduced by only 6 percent, equivalent to 1 percent every 14 years. Between 1960 and 1982, that rate increased to 1.6 percent, and from 1982 to 1991, it reached a rate of 3.5 percent per year (Fuentes Aguilar and Soto Mora 1992).

Subregional studies showed that approximately 14,700 hectares a year were destroyed along the Rio Usumacinta between 1980 and 1988 (Cortez-Ortiz 1990, cited in Masera 1996). Based on an analysis of satellite imagery, Buerkle (1996) estimated that 30.1 percent of the Marqués de Comillas region was lost by 1989, with all but 2.7 percent of the deforestation occurring after 1979. This amounts to annual deforestation rates of 2.8 percent during a ten year period. Between 1989 and 1993, an additional 8.1 percent was deforested, such that by 1993, 38.2 percent was deforested, equivalent to approximately 78,000 of the 204,000 hectares analyzed.

Estimates of deforestation rates in the Selva Lacandona are contingent upon the size of the area or region under consideration. The "Selva Lacandona" can refer to a number of different regions, defined according to various criteria. These criteria include municipal or political boundaries and biogeographic boundaries, including the current and historical forest borders. References to the social aspects of deforestation tend to be more inclusive, taking into account the highland areas with coniferous forests, as well as the long-ago deforested areas around Palenque. From a biological perspective, the limits of the Selva Lacandona are usually restricted to areas with tropical forest vegetation.

Satellite Views of Deforestation

Many estimates of deforestation suggest that over half of the forest cover has been destroyed in the Selva Lacandona. However, because most of these estimates are confined to an area containing tropical rain forest vegetation, they do not provide a reliable picture of deforestation in the larger region. To gain this wider perspective, satellite images were used to evaluate changes in forest cover in the Selva Lacandona region.

For the purpose of understanding the driving forces of deforestation, a broad boundary was used to define the Selva Lacandona region. The Selva Lacandona encompasses an estimated 1.89 million hectares, located between 16°05' and 17°30' N. latitude and 90°25' and 92°15' W. longitude. This corresponds to the Chiapas portion of the Usumacinta river basin, which contains both coniferous and tropical forests, rivers, lakes, and other non-forest vegetation and geographic features. From a political or administrative perspective, the Selva Lacandona includes the municipalities of Ocosingo, Las Margaritas, and Altamirano, as well as small parts of Palenque, Chilón, La Independencia, and La Trinitaria (Figure 3.1).[5] However, it is important to emphasize that not all communities within the three main municipios share the characteristics of populations of the Selva region. For example, some communities in Altamirano are more closely identified with the Highlands than with the Selva (Leyva Solano and Ascencio Franco 1996).

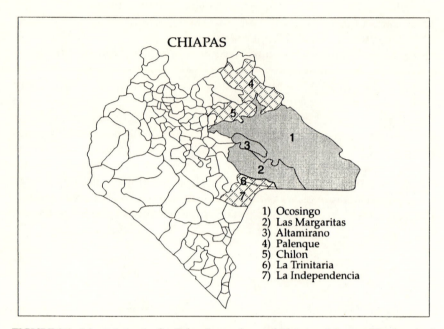

FIGURE 3.1 Municipios in the Selva Lacandona of Chiapas, Mexico. The hatched shadings indicate municipios where only a small portion of the land is located in the Selva Lacandona region.

This broadly-defined region includes a wide swath of farms located along the western and northern perimeter. These lands were among the first areas to be settled, and today contain mere fragments of forests. The region also includes the Lagos de Montebello and surrounding areas, located just north of the Guatemalan border on the western edge of the Selva Lacandona region. The pine-oak forests in this area serve as a port of entry into the southernmost forests of Las Margaritas.

To estimate the extent of deforestation in the Selva Lacandona, satellite images representing two time periods were classified, then mosaicked to produce a continuous coverage for the region. Data sources and details of the classification procedure and deforestation analysis are presented by O'Brien (1995).[6] A composite image representing the period from 1974 to 1979 is shown in Figure 3.2, and an identical coverage representing the period from 1989 to 1992 is shown in Figure 3.3. The reduced size of the images hides much of the detail, but can provide a general overview of deforestation in the region. Patterns of deforestation in subregions of the Selva Lacandona are described below.

In Figure 3.2, an estimated 78 percent of the region is covered with forest or natural vegetation, and 10 percent is represented by cleared areas. The remaining 12 percent of the image is obscured by clouds, cloud shadows, or terrain shadows. Water is also included in the uncertain category, as it could not be spectrally separated from terrain shadows in all of the satellite images. One means of estimating the land cover under these "uncertain" areas is to assign the pixels to categories based on the ratio between known forested to cleared areas. Given the higher proportion of forest relative to cleared areas, this method may be biased toward forest cover.

According to this method, it is estimated that 90 percent of the Selva Lacandona was still forested in the 1974-1979 period (Table 3.1). This is a very conservative estimate of forest cover, given that secondary forest and tree crops such as cacao are included with the forest category. However, it is also true that the earliest settlers in the forest did not clear more than a few hectares in their first years, only enough to establish small *milpas*.[7]

In Figure 3.3, about 60 percent of the area is covered with forest, while 21 percent is characterized as cleared. Nineteen percent of the area could not be classified with certainty. If this area is distributed according to the ratio of forested to cleared land, then 14 percent of the uncertain area can be considered forest and 5 percent cleared land. The resulting approximation indicates that 74 percent of the Selva Lacandona was covered with forest in the period between 1989 and 1992. This means that of the 1.89 million hectares of original forest in the Selva Lacandona region, approximately 492,000 hectares, or 26 percent, was deforested by the early 1990s.

The estimates of forest cover were used in calculating the actual forest change between the two periods, and the result was a deforestation of approximately 2,775 square kilometers, or 277,500 hectares between the two periods. If the total forest is considered to be 1.89 million hectares, then the area deforested between 1974-1979 and 1989-1992 is equivalent to 15 percent of the total Selva Lacandona, tantamount to 87 percent of the area that is currently preserved in the Montes Azules Biosphere Reserve.

The calculations suggest that 56 percent of the forest loss occurred in the period between 1974-1979 and 1989-1992, in comparison to 44 percent in the period prior to 1974. Since the Selva Lacandona remained largely forested as late as 1960, most of the 44 percent was probably deforested during the 15-year period between 1960 and 1974.

It is worth reiterating that rivers and water bodies are considered part of the uncertain category in this classification, thus their total

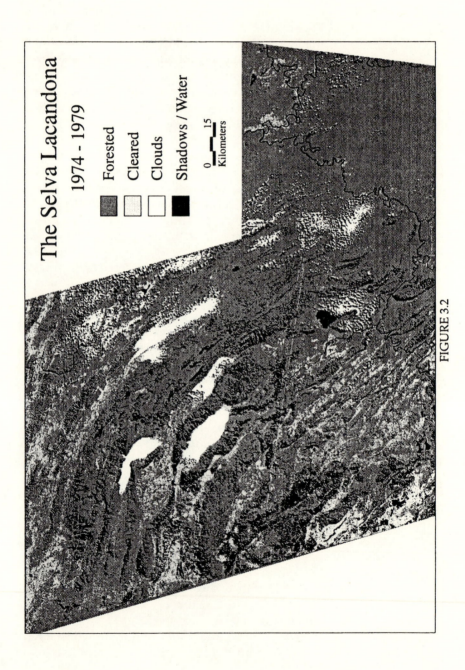

The Selva Lacandona
1974 - 1979

Forested
Cleared
Clouds
Shadows / Water

0   15
Kilometers

FIGURE 3.2

The Selva Lacandona
1989 - 1992

Forested
Cleared
Clouds
Shadows / Water

0   15
Kilometers

FIGURE 3.3

TABLE 3.1 Estimates of deforestation in the Selva Lacandona,
based on the interpretation of Landsat imagery.

Category	1974-1979		1989-1992	
	(km ^{2})	*Percent*	*(km ^{2})*	*Percent*
Forested	14 779	78	11 352	60
Cleared	1 850	10	3 945	21
Uncertain	2 295	12	3 623	19
Estimated	**2 105**	**11**	**4 879**	**26**

Source : O'Brien, 1995.

surface area is distributed among forested and cleared land according to
the ratio described above, adding inaccuracy to the estimate. The
importance of secondary growth in agricultural systems is also under-
stated by these estimates, given that secondary growth is treated as
forest. Given these limitations, one may conclude that *at least* 26
percent of the forest had been lost by the early 1990s, as represented by
the 1989-1992 mosaicked image.

The pace of deforestation has become increasingly rapid over the
past five years, in part due to a momentum created by colonists that
have established themselves in the region, and in part due to the
increasing integration of the region with the outside world through new
or improved roads. Considering field observations made between 1992
and 1996, it is reasonable to estimate that about 40 percent of the Selva
Lacandona had been disturbed or cleared by 1995. This amounts to about
756,000 hectares of forest destroyed. A substantial part of the remaining forest consists of coniferous trees located at higher elevations in the
western part of the region. The remaining tropical forest is becoming
more and more isolated along ridges, in fragmented patches, and within
protected areas.

Patterns of Deforestation

The evaluation of deforestation in the Selva Lacandona region
presented above indicates that a smaller area has been destroyed than
suggested by many previous estimates. The discrepancy can be attributed to differences in the methods of evaluating deforestation, and to
the different boundaries used to define the region. The conservative
nature of the classification presented here must also be acknowledged.

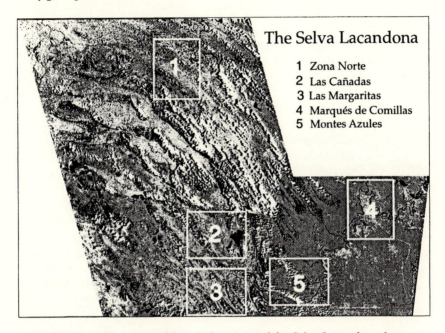

FIGURE 3.4 Areas extracted from subregions of the Selva Lacandona for more detailed analysis of deforestation.

Nevertheless, it is important to stress is that deforestation is a heterogeneous process, and not all areas of the Selva Lacandona have experienced the same amount of forest loss. The patterns of forest loss also vary across subregions. Whereas some parts of the Selva Lacandona have experienced 70 or 80 percent deforestation, other parts have experienced virtually no changes in forest cover. In between these extremes lie a variety of patterns that are closely related to the driving forces of deforestation that will be discussed in subsequent chapters.

To illustrate the diversity of deforestation patterns across the region, five subsets of the mosaicked images have been extracted for comparison (Figure 3.4). The subsets, each 72,000 hectares in size, correspond to five of the six regions identified by Leyva Solano and Ascencio Franco (1996) based on their fieldwork in the region: (1) the Zona Norte; (2) Las Cañadas de Ocosingo-Altamirano; (3) Las Cañadas de Las Margaritas; (4) the Marqués de Comillas region; and (5) the Montes Azules Biosphere Reserve. The sixth subregion corresponds to the Comunidad Lacandona in the east, and was not extracted due to the high percentage of cloud cover in both images.

The first part of the Selva Lacandona to experience heavy deforestation was the northern region, referred to here as the Zona Norte. Much of the forest in this subregion was converted to cattle ranches at the end of the early logging period, particularly in the municipio of Palenque. The remaining forests were among the first areas to be heavily logged by commercial loggers in the 1960s, and colonized. Although logging roads made many areas accessible for colonization, spontaneous settlements emerged prior to the existence of roads.

The area extracted to represent the Zona Norte lies just west of the Río Usumacinta, and about 5 kilometers north of the Montes Azules Biosphere Reserve. This area is part of the Santo Domingo valley, which contains the Río Santo Domingo, running from the northwest to southeast. Three lakes, Lagunas Metzabok, Baquelté, and Guinea, are located in the southwest, and the Río Chocoljá runs across the northeast corner of the image (Figure 3.5). The communities in this area represent some of the oldest settlements in the Selva Lacandona.

In the 1974 image, deforestation is concentrated along the Santo Domingo valley. Although some settlements can be distinguished by a large nucleus of clearings, often the clearings of one community merge into those of another. In addition to settlements along the Santo Domingo valley, a number of clearings can be seen along the Río Chocoljá to the north. Small, nucleated settlements are also evident in the parallel valleys. Small clearings are visible in and along the mountains, where many families maintain milpas. Although a significant amount of land had been cleared, much of the area was still forested in 1974.

Dramatic changes took place in this area over the following two decades. By 1992, most of the Santo Domingo valley had been cleared, and only small patches of forests remained. In the valley to the north, a new highway was completed in the early 1980s. This road, extending from Palenque in the northwest to Benemérito de las Americas in the southeast, has clearly influenced the patterns of deforestation in this area. By 1992, this subset contained over 80 settlements, ranging from isolated ranches to small groupings of houses or farms, as well as larger villages (Ecosur 1997). Some of the larger settlements, including Arroyo Granizo, Ubilio García, Jerico, Nuevo Jerusalén, Damasco, and San José Patihuitz, are marked on the image. The populations of selected ejidos in 1990 were as follows: Arroyo Granizo (883); Ubilio García (731); Jerico (488); Nueva Jerusalén (418); Damasco (1,554); San José Patihuitz (773); Nuevo Canán (365); Once de Julio (119); and Ricardo Flores Magon (214) (INEGI 1991).

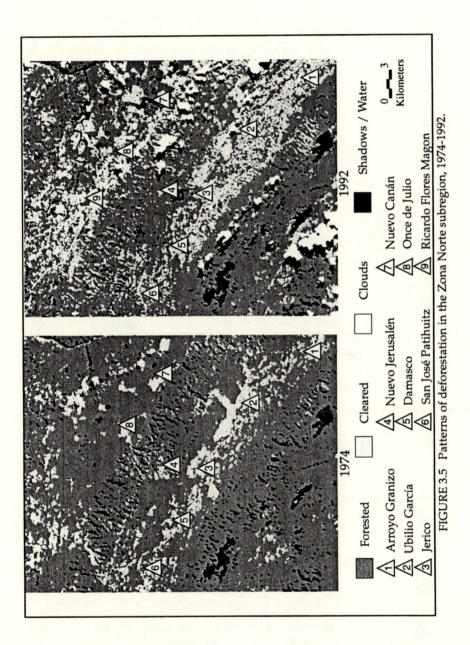

FIGURE 3.5   Patterns of deforestation in the Zona Norte subregion, 1974-1992.

The two images presented in Figure 3.5 illustrate how patterns of deforestation evolved in many valleys of the Selva Lacandona. Colonized initially by a few communities, the area under cultivation or in pasture expanded to occupy a larger and larger proportion of the area. For example, Arroyo Granizo was founded in 1958, largely by Tzeltal Indians from various municipios of Chiapas. Many of the colonizers were wage laborers in search of land, while others had left behind degraded lands in their villages of origin (S.R.H. 1976). These settlers cultivated maize, beans, squash, and fruit trees for subsistence, often selling the surplus to itinerant buyers or *coyotes*. As in many communities, once the most productive lands became scarce, or when harvests decreased, younger members would have to sell their labor on nearby farms or seek employment in distant cities. In many cases, younger members simply ventured to other areas in search of more fertile lands. Another pattern common to early settlements are the satellite clearings that emerged when individual families purchased lands and establish small private ranches away from the ejido center. These satellites are culturally and socially tied to ejidos, but independent in terms of production activities (Leyva Solano and Ascencio Franco 1996).

Somewhat similar patterns of deforestation can be seen in Las Cañadas subregion. The area represented in Figure 3.6 is located in the municipio of Ocosingo, and includes the San Quintín valley along the lower Río Jataté, just to the west of Laguna Miramar. The Río Perlas feeds into the Río Jataté from the north. Further to the west runs the Río Euseba, just on the other side of the Sierra la Colmena. This subset represents one of the last colonization frontiers before the Montes Azules Biosphere Reserve. In fact, conflicts have already arisen over settlements established within the Montes Azules Biosphere Reserve, located in the eastern third of this subset. Land tenure conflicts between ejidos and the Comunidad Lacandona have also emerged in the northeastern part of this area.

Colonization of this area began in the 1960s, and accelerated during the 1970s and 1980s. A number of settlements were well-established by 1974. Most of the early settlements were located along rivers. For example, the village of San Quintín, located on the Río Jataté near the mouth of the Río Perlas, was established in 1965 at an old logging camp of the Casa Bulnes Hermanos. The large cleared area across the river from San Quintín is a natural savanna, covered with grasses and sparse trees. San Quintín is largely a tzeltal community, with some Tojolabal Indians from Las Margaritas. With a population of 812 in 1990, it forms the largest settlement in this area of 72,000 hectares.

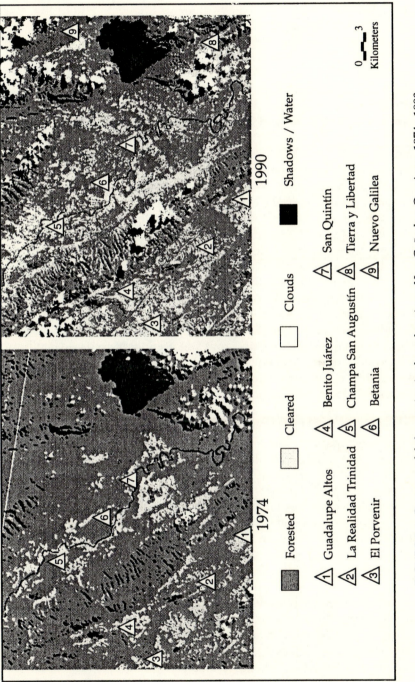

1974

1990

Forested · Cleared · Clouds · Shadows / Water

San Quintín

△1 Guadalupe Altos △4 Benito Juárez △7 Champa San Augustín △8 Tierra y Libertad

△2 La Realidad Trinidad △5 Champa San Augustín △9 Nuevo Galilea

△3 El Porvenir △6 Betania

0 ⎯ 3
Kilometers

FIGURE 3.6 Patterns of deforestation in the subregion of Las Cañadas, Ocosingo, 1974 - 1990.

By 1990, there were over 65 additional settlements within this area (Ecosur 1997). The valley floors have been largely deforested, and colonization within the Montes Azules Biosphere Reserve is clearly evident. There is also substantial evidence of clearings along the slopes of the Sierra la Colmena. The population of selected settlements include the following: Guadalupe Altos (340); La Realidad Trinidad (318); El Porvenir (145); Benito Juárez (160); Champa San Augustín; Tierra y Libertad (108); and Nuevo Galilea (219) (INEGI 1991).

The patterns of colonization evident in this subset are similar to patterns in the Zona Norte. However, access to this area is much more difficult, and the settlements are consequently more isolated from regional markets, and lacking in infrastructure and services. This subregion comprises one of the core areas of support for the Zapatistas. The process of colonization in the San Quintín valley is described in detail by Leyva Solano and Ascencio Franco (1996). They emphasize the links between communities, formed when the descendants or relatives of one ejido or village move away to establish their own colony, from which new ejidos or ranches are later formed. For example, the founders of Champa San Augustín arrived from a private farm called La Martinica in the mid-1960s to settle along the Río Jataté. A few years later, relatives followed and established the ejido Betania further down the river. Eventually, descendants from these ejidos purchased land and established separate ranches. In highlighting the linkages between settlements in this area, Leyva Solano and Ascencio Franco (1996) demonstrate how microregional economies have emerged within some parts of the Selva Lacandona region.

Directly south of Las Cañadas in Ocosingo lies the subregion referred to as Las Margaritas. The extracted area is centered on the Río Santo Domingo and several tributaries that feed into it from the northwest (Figure 3.7). The terrain of the region consists of parallel mountain ridges divided by fertile valleys. The southernmost portion of the image includes part of Guatemala. The tropical forests of Guatemala were colonized earlier than the Selva Lacandona, and distinct differences in land cover can be seen in the 1974 image.

While colonization of the Zona Norte and some parts of Las Cañadas got underway in the 1950s, it was not until the 1960s that Las Margaritas was opened up for colonization. By 1974, a number of settlements had been carved out of the valleys of this subregion, particularly in the northwest. By 1990, most of the forests along these valleys had been cleared and replaced with cropland or pastures. Forest cover was increasingly restricted to mountain ridges, and even some of those forests were being cleared for agriculture.

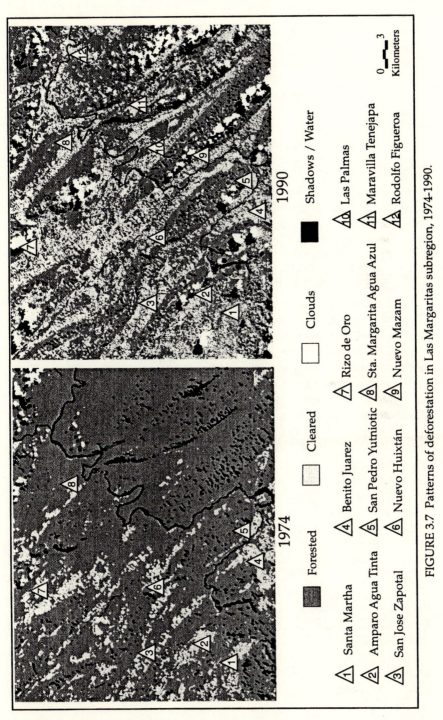

FIGURE 3.7 Patterns of deforestation in Las Margaritas subregion, 1974–1990.

Some of the earliest settlements in this area were established by migrants from the Highlands of Chiapas. In almost all cases, new settlers were searching for land to cultivate. Many of the names of ejidos or communities reflect the origins of the settlers. For example, Nuevo San Juan Chamula was settled in 1964 by migrants from Chamula, and Nuevo Huixtán was founded in 1965 by migrants from the municipio of Huixtán (Paz Salinas 1989). Between 1964 and 1975, at least 10 settlements were established by migrants from the Highlands, and at least one dozen more were formed by colonists from other parts of Chiapas. By 1990, there were over 125 settlements within the 72,000 hectares of the subset, with colonization extending as far east as the Río Lacantún. The populations of selected settlements had grown considerably by that time: Santa Martha (64); Amparo Agua Tinta (490); Benito Juárez (339); San Pedro Yutniotic (593); Nuevo Huixtán (445); Rizo de Oro (294); Santa Margarita Agua Azul (312); Nuevo Mazam (337); Las Palmas (514); Maravilla Tenejapa (176); and Rodolfo Figueroa (375).

Unlike the situation in some parts of the Zona Norte and Las Cañadas subregions, there were few conflicts over land in this area. Most of Las Margaritas subregion consisted of unclaimed national lands that were open for colonization. There were neither biosphere reserves in this area, nor communal lands belonging to a particular Indian group, such as the Comunidad Lacandona. Nevertheless, Paz Salinas (1989:90) points out that colonization of this region did not take place in a homogeneous or equitable manner. The earliest settlers claimed the best land, generally receiving larger and better parcels than later arrivals. Proximity to roads and rivers also held advantages, and many of the later colonizers were clearly at a disadvantage. With the pavement of the road extending eastward from the Lagos de Montebello in 1994, many communities within this subregion became more easily accessible. Crops such as coffee, which were earlier difficult to market, became viable alternatives for production. With the completion of the improved road, a bus ride from Maravilla Tenejapa to Comitán was reduced from 7 hours to less than 3 hours.

The area selected to represent the Marqués de Comillas region (Figure 3.8) is located in the easternmost part of the forest, at the confluence of the Ríos Lacantún and Salinas (Chixoy), which together form the Río Usumacinta. In the 1979 image, only a small portion of this area had been cleared of tropical forest. Most of the deforestation was, in fact, in Guatemala, on the right banks of the Salinas (Chixoy) and Usumacinta rivers. Small clearings are evident on the Mexican side, particularly along the rivers. Some of these clearings represent

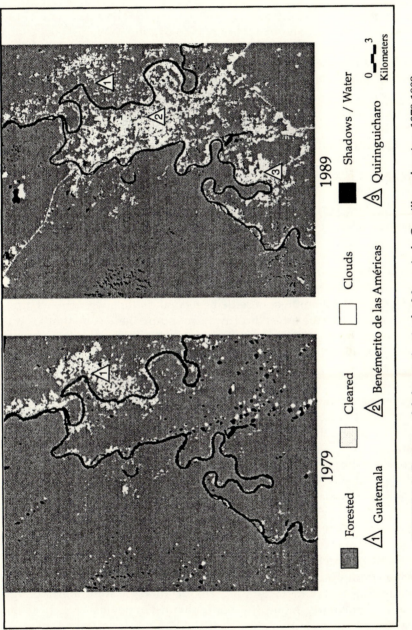

FIGURE 3.8 Patterns of deforestation in the Marqués de Comillas subregion, 1979-1989.

land cultivated by ejidos, while a smaller proportion represent natural clearings vegetated with bamboo-like grasses known as *jimbales*.

By 1979, several ejidos had been established in this area, including Benemérito de las Américas and Quiringuicharo. The small sizes of the areas cleared reflect the dynamics of colonization. Although the ejidos were established by presidential resolution in 1963 for Quiringuicharo and in 1965 for Benemérito, colonization of these *Nuevo Centros de Poblaciones Ejidales* (NCPEs) was a long and challenging process.[8] Many of the earliest colonizers left the forest, and it often took time to find others to replace them. Ten years later, visible changes had taken place within this small area. By 1989, the communities of Benemérito and Quiringuicharo had both expanded considerably. The population of the former was estimated to be 3,320 in 1990, with an additional 72 people in a northern annex, officially referred to as Benémerito Section II (INEGI 1991). The population of Quiringuicharo in the same year was 712. The road to Palenque, seen in the upper left part of the image, was completed in 1984. Traces of its extension from Benemérito south to the PEMEX facilities and continuing to the Guatemalan border can be seen in the image. A road heading southeast to Quiringuicharo is also visible. These roads had enormous impacts on the social and economic lives of the inhabitants of the Marqués de Comillas region. Between 1985 and 1992, the number of settlements in the subregion doubled, from 19 to 38. Much of the forest directly surrounding these communities was fragmented by 1989, and all but remnants remain today.

The situation on the Guatemalan side of the river is quite different. The settlements visible in the 1979 image all but disappeared by 1989, an outcome of the civil war and the Guatemalan army's brutal counterinsurgency tactics. Many of the inhabitants of this area fled across the river to Mexico from 1980 to 1983, abandoning farms and taking refuge in the dozens of camps setup throughout the Selva Lacandona. By 1996, many of these areas were repopulated, and the secondary growth forests that had regenerated were again being cut.

The patterns of deforestation that characterize the Marqués de Comillas region are more nucleated than those in many other parts of the forest. This can be attributed to the larger size of the land distributions, which has enabled many ejidos and communities to maintain large areas in forest reserves. It is only along the rivers, where some of the most productive lands are located, that clearings often merge to form a continuous band of deforestation. Although roads have played an important part in facilitating colonization in some parts of the region, particularly along the border with Guatemala, much of the area lacked roads until very recently. People instead relied on river transport to facilitate movement within the region and foot, horseback, or sometimes airplanes to travel outside of the region. The expansion of road construction in the past few years has altered the character of the area, from a remote frontier to an accessible and dynamic region.

The final subregion of the Selva Lacandona considered here is the Montes Azules Biosphere Reserve, much of which coincides with land belonging to the Comunidad Lacandona. The area extracted (Figure 3.9) includes the southern part of the reserve, directly across from the Marqués de Comillas region discussed above. The Montes Azules Biosphere Reserve is located to the west of the Río Lacantún in this image. In the 1979 image, virtually none of the forest has been cleared on the Montes Azules side. The very small patches of clearing represent a fern-like vegetation that grows naturally on specific soils.

By 1979, only small patches of forest had been cleared along the southern side of the Río Lacantún. The community of Pico de Oro, one of the earliest settlements in the Marqués de Comillas region, can be identified in the 1979 image. Some small clearings made by early settlers can also be seen along the southern side of the river.

One decade later, some notable changes can be seen on the Marqués de Comillas side of the Río Lacantún. By 1989, the ejidos of Reforma Agraria, Adolfo López Mateos, Galaxia, and Playón de la Gloria had been firmly established. These ejidos had populations of 171, 272, 144, and 186 in 1990, respectively. Nevertheless, the forests within the Montes Azules Biosphere remained largely intact, at least along its southern border. Ironically, the lack of clearings in this area cannot be attributed to the existence of the biosphere reserve. Colonizers were hardly aware of the decree, and there was little vigilance or institutional presence in the area prior to 1989. The lack of clearings can instead be attributed to the fact that the limited number of ejidos established in the Marqués de Comillas region received ample land to cultivate on the other side of the river from the reserve.

An analysis of deforestation in these five subregions shows that the amount of forest loss varies substantially within the Selva Lacandona region. Likewise, the patterns of deforestation vary among subregions, depending on a number of factors, including both external pressures and internal dynamics. What becomes clear from this brief analysis is that "deforestation in the Selva Lacandona" is not a process that can be easily generalized. Each settlement exists within both historical and locational contexts. These contexts were shaped by extractive activities, agricultural transformations, agrarian politics, political upheavals, and conservation policies, and are discussed in the following chapters.

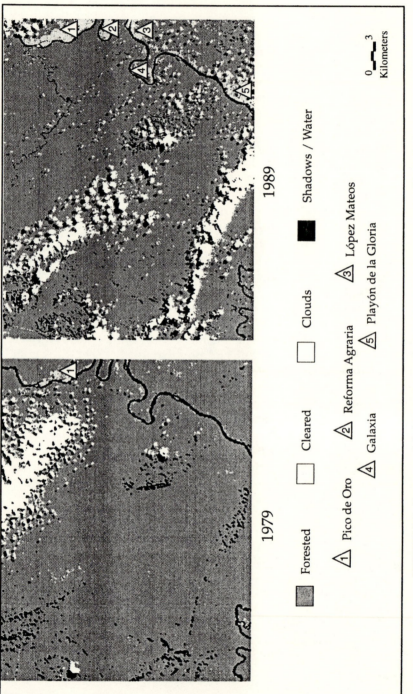

1979   **1989**

Forested   Cleared   Clouds   Shadows / Water

△1 Pico de Oro   △2 Reforma Agraria   △3 López Mateos

△4 Galaxia   △5 Playón de la Gloria

0 ___ 3
Kilometers

FIGURE 3.9   Patterns of deforestation in the Montes Azules subregion, 1979-1989.

4

Facilitating Access

Extractive Industries

Until the emergence of widespread concern over the loss of tropical forests, the Selva Lacandona was valued by many only for the economic rewards that could be extracted and sold in the marketplace. Extractive industries, concentrated in areas where resources are both abundant and accessible, have indeed been the most important production activity in the Selva Lacandona throughout much of its modern history. The logging of tropical hardwoods and the extraction of products such as petroleum, chiclé, and xaté palm have become important activities during different periods, with different impacts on the forest.[1]

The logging of tropical timber from the Selva Lacandona is widely believed to have initiated the deforestation process in what is often erroneously referred to as virgin forest.[2] More recently, petroleum exploration has also been indicted in the destruction of the rain forest. Both industries are singled out as direct causes of deforestation for the removal of trees associated with logging and drilling, as well as for the construction of roads that facilitate access to the resources in remote parts of the forest.

To condemn extractive industries as destructive *per se* is to grossly oversimplify the situation. If extractive industries are examined from the perspective of political ecology, that is, in relation to issues of land, labor, and production relations, then different conclusions emerge. It becomes evident that extractive industries are not necessarily destructive in and of themselves. Instead, the impacts are closely related to the ways that the activities are carried out. In the case of logging in the Selva Lacandona, the state has consistently intervened to favor large companies or state-run enterprises, while at the same time excluding local communities from developing one of the few potentially profitable and sustainable production activities.

This chapter explores how extractive industries have evolved in the Selva Lacandona, and how the impacts on deforestation have varied across time and space. The evolution of timber extraction is traced, from the "opening" of the Selva Lacandona in 1860 to the establishment of private, foreign-financed logging companies, followed by a period of decline in the second half of the twentieth century and a transfer to state-control. The logging ban and the recent and controversial distribution of logging rights to communities within the Marqués de Comillas region of the Selva Lacandona is also considered. In addition, petroleum exploration is examined, as it provides a modern example of the complex relationships between state-sponsored resource extraction, community development, and environmental protection.

Logging in the Selva Lacandona

Logging has been the dominant extractive activity in the Selva Lacandona for over 100 years. Like other industries, the logging era has been marked by peaks and declines corresponding to changes in land use regulations, labor conditions, production processes, and external politics. To understand the relationship between logging and deforestation, it is necessary to consider both the development and decline of the logging industry in the Selva Lacandona. It is also worth considering options and obstacles to developing community forestry in the region, as this is viewed by many as a potentially sustainable use of the forest. Such considerations involve an examination of land acquisition policies and procedures, the operational components of logging, and forest politics in Chiapas.

The earliest logging in the Selva Lacandona was highly selective, focusing on mahogany and tropical cedar. It can be argued that such selective logging does not constitute deforestation at all, as it does not involve the total clearance of forest. Instead, it can be considered a form of land degradation (Grainger 1993). Based on his extensive history of the Selva Lacandona, de Vos (1992) concluded that the environmental impacts of early logging were minimal, largely because it was not capital intensive. Early logging ventures were concentrated along the region's expansive network of rivers (Figure 4.1). Although the establishment of logging camps within the forest and the haciendas that provided for them can be considered the earliest phase of modern

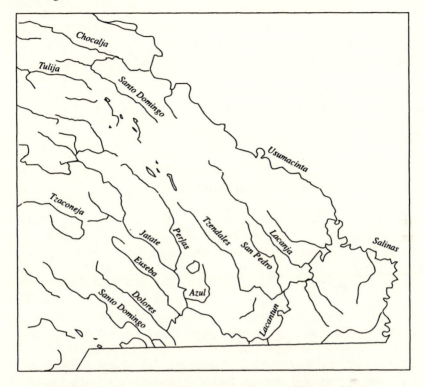

FIGURE 4.1 Rivers in the Usumacinta basin of Chiapas, Mexico.

colonization, the impacts of logging were constrained by existing technology and difficult access to the region.

It was not until production techniques became increasingly mechanized that the logging industry could be associated with extensive forest destruction. Modern logging not only penetrated larger areas of forest with bulldozers and chainsaws, but also harvested common tree species that had been of no interest to earlier companies. As modern technology was increasingly incorporated in the production system, logging resulted in greater direct damages to the forest. Furthermore, it created externalities which have had a far more devastating impact on the forest. The externality of greatest significance has been the network of roads constructed to access new stands of timber. These roads have facilitated colonization of once-inaccessible areas, which has in turn resulted in increased deforestation.

The story of logging in the Selva Lacandona does not, however, end with the large-scale destruction of the forest associated with mechanized production processes. This type of logging, concentrated in specific parts of the forest, ceased to be profitable for a number of reasons that will be discussed below. Alternatives to commercial logging, carried out by communities on their own forest lands, were rejected in favor of a complete ban on logging in the name of conservation. This decision not only excluded residents of the region from a potential source of income that could be achieved on a sustainable basis, but also exacerbated tensions between social and environmental struggles in the Selva Lacandona.

Land Acquisition

To understand who had access to land in the Selva Lacandona during the various logging periods, and how it was obtained and used, one must go back to the middle of the nineteenth century. Prior to 1863, private property did not exist in the Selva Lacandona (González Pacheco 1983). Rules for claiming property were first established in 1863 under the *Ley sobre Ocupación y Enajenación de Terrenos Baldíos*, signed by Mexican president Benito Juárez. This law, disguised as an agrarian reform, established the rules for titling "vacant lands" (*terrenos baldíos*). Much of the 11.1 million hectares that entered into the hands of *latifundistas*, or large landholders, in the years following the enactment of this law belonged to Indians who allegedly could not prove title to their land (Cockcroft 1990). As a result, *criollo* and *mestizo* landholders vastly increased their holdings at the expense of Indian groups throughout Mexico. This occurred in spite of the fact that individual claims were limited to no more than 2,500 hectares, as titles were merely placed in the names of relatives or nonexistent persons.

The 1863 law also had several stipulations tied to it which discouraged land accumulation by small-holders or landless persons. First, no land could be claimed unless it had been professionally surveyed, with the claimant paying all necessary costs. Second, the land had to be purchased, in part to finance the national government's fight against French invaders (Cockcroft 1990). The price for the land was set by the government and updated every two years. Finally, the entire sum had to be paid at the time that the title was handed over (de Vos 1988a:68). This made it extremely easy for latifundistas to amass even larger landholdings, yet practically impossible for small-scale farmers to do the same.

Landholdings were further consolidated during the political rule of Porfirio Díaz. During the period between 1876 and 1910, often referred to as the *Porfiriato*, Mexico was opened up to foreign investment and became an integral part of the North American economy. A series of national laws were passed by Díaz, aimed at promoting agricultural development, establishing land registries, privatizing agricultural lands, and colonizing unsettled territory (Gutelman 1971). These laws made it yet easier for latifundistas, including foreigners, to acquire large tracts of land in Mexico.

In 1883, the *Ley de Colonización* was passed with the objective of opening up pristine areas of Mexican territory. This law gave surveying companies the right to keep up to one-third of the land surveyed as compensation for expenses incurred, and to provide an incentive to survey unexplored territory. The same companies were given the option to purchase the remaining two-thirds of the land, and were offered preferential tariffs as an incentive. The official titles were not to exceed 2,500 hectares, but again, this was rarely adhered to. By 1889, surveying companies in Mexico had acquired 27.5 million hectares, or 13 percent of Mexican territory (Gutelman 1971).

In 1893, the Mexican Congress passed a new law which amplified the 1883 legislation. The new *Ley sobre Ocupación y Enajenación de Terrenos Baldíos* replaced all previous legislation concerning vacant or unused lands. Surveying companies were still entitled to keep one-third of the land surveyed, but titles were no longer limited to 2,500 hectares. Furthermore, stipulations that the land had to be colonized and fenced in were removed (Gutelman 1971). Indian lands which had no legal titles attached to them could also be incorporated into the newly surveyed lands. Land could be denounced by individuals as well as companies, which paved the way for land speculation.

To encourage surveying companies to venture into remote or less-than-desirable areas such as the Selva Lacandona, the law included an option for procuring yield contracts. Those holding such concessions would not have official title to the land, but could profit from the wood, resins, or other products associated with the land. Laws and regulations passed in subsequent years made it increasingly less complicated for logging companies to gain title to land on which they held concessions (de Vos 1988a).

In Chiapas, the obvious result of such laws was that many commercial enterprises, including logging companies, transformed themselves quickly into surveying companies and acquired government contracts to survey and colonize large tracts of land. During the Porfiriato, Chiapas landowners amassed huge landholdings and ventured into large-scale export production of sugar cane, cotton, indigo, rubber, and cattle, as well as tropical timbers. Between 1875 and 1910, titles to land in the Selva Lacandona passed between speculators in

Mexico City, foreign owners, and timber companies who were eager to expand into the untapped forests of Chiapas. Tabascan and foreign logging companies in particular were successful in establishing themselves in the region, not only through property holdings, but also through government contracts. Their presence clearly dominated the early history of the Selva Lacandona (González Pacheco 1983; de Vos 1988a).

It is worthwhile to examine the history of these land accumulations in more detail, as they illustrate the political manipulations characteristic to the formation of large estates in the Selva Lacandona. One of the biggest logging firms to establish a presence in the Selva Lacandona was the Casa Bulnes Hermanos (de Vos 1988a; González Pacheco 1983). Between 1880 and 1926, Casa Bulnes extracted an estimated one million tons of valuable hardwoods from the forest, with a commercial value of between 50 and 70 million pesos (Benjamin 1981). The two Bulnes brothers, Antonino and Canuto, emigrated to Mexico from Spain in 1855 to join an uncle who had settled in Tabasco. They married two local women, and were able to establish themselves on some land near Pichucalco, Tabasco. The Bulnes brothers eventually developed a successful monopoly in the transportation sector in Tabasco, and consequently acquired sufficient capital to venture into the logging industry (de Vos 1988a). The logging industry in Tabasco, however, was already firmly under the control of wealthy tabasqueños, including Policarpo Valenzuela and Manuel Jamet, as well as several Spanish families such as Romano, Ramos, and González. With little prospect of setting up successful logging operations in Tabasco, the Bulnes brothers set sights on the Jataté river basin in neighboring Chiapas.

The Bulnes operations prospered and came to dominate the logging industry in the western part of the Selva Lacandona for over 30 years. Prior to its demise following World War I, the company held land assets consisting of 22 properties or 48,113 hectares (González Pacheco 1983). Furthermore, the business transactions of the logging enterprises were extremely profitable. González Pacheco (1983) examined the 1914-1915 financial records of the Casa Bulnes and found that, even in the declining years, profits made up 20 percent of the total balance. Of these profits, four members of the Bulnes family received over 50 percent, whereas the work force received a mere 0.02 percent of the total balance.

Despite the seemingly low operating costs associated with an industry that relied heavily on an exploited labor force, a considerable amount of money had to be paid out to the state and federal government for property titles, surveying contracts, and logging rights. Fees also had to be paid to engineers for mapping and surveying the land, and to lawyers in Mexico City and San Cristóbal for cutting through bureaucracy (de Vos 1988a). This money was almost always financed by international capitalists. González Pacheco (1983) traces the penetration of foreign capital into the Selva Lacandona, as well as the history of the local timber companies that were dependent upon it.

The history of land accumulation by other timber barons, such as Policarpo Valenzuela and the Romano family, follows similar patterns. Valenzuela, a powerful capitalist from Tabasco, was the first to exploit timber in the forests of Chiapas. Although his landholdings in southeast Mexico totalled over one million hectares, he was also involved in shipping and transportation, petroleum exploration, as well as the banking sector (González Pacheco 1983). In the Selva Lacandona, *Compañía Valenzuela* obtained title to over 100,000 hectares along the Usumacinta river, including land where the Maya ruins of Yaxchilán were located. The Valenzuela property was situated in the eastern part of the Selva, distant from the operations of Casa Bulnes Hermanos. Consequently, these two firms came to co-dominate the timber trade in the Selva Lacandona. A third major logging company, *Compañía Romano y Sucesores*, was controlled by Spaniards with extensive experience in logging the forests of Tabasco. Through the acquisition of logging concessions, surveying compensation, and outright purchases, the Casa Romano was able to gain control of large areas of land within the Selva Lacandona, concentrated primarily along the Río Tzendales (de Vos 1988a).

The distribution of logging concessions within the Selva Lacandona can be seen in Figure 4.2. This map shows how large tracts of the forest were divided up among nine logging companies between 1897 and 1900. Most of the concessions were located along the Lacantún and Usumacinta rivers, whereas the central part of the Selva remained untouched by the logging industry due to its inaccessibility and the lack of a river-based transportation system for extracted timber. The western part of the forest, along the Río Santo Domingo, was also not heavily logged, either for lack of valuable timber species or for difficulty transporting the timber out of the region (Paz Salinas 1989).

The history of land access and accumulation by timber companies in the Selva Lacandona would be incomplete without mention of the role that land speculators played in dividing up the forest. Many of the speculators were from Mexico City, and most of them had close connections to powerful politicians. One of the earliest successful land speculators, Rafael Dorantes, was a personal friend of Porfirio Díaz. In 1892, he acquired a contract of sale and colonization for 300,000 hectares

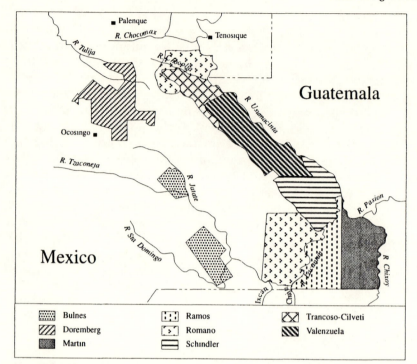

FIGURE 4.2 Logging concessions in the Selva Lacandona; 1897-1900 (Adapted from de Vos, 1992).

of national land in Tabasco and Chiapas. This eventually led to the entitlement in 1896 of over 13,000 hectares in the municipio of Palenque (de Vos 1988a). These acquisitions were followed by others, such that by 1911, Dorantes had accumulated possessions of over 241,000 hectares in Chiapas, as well as large holdings in Tabasco and Campeche. These lands were flat and particularly amenable to rubber cultivation.

Another friend of President Díaz, Luis Martínez de Castro, acquired title to over 323,000 hectares in the central part of the Selva through the same process, as did Antenor Sala, Julio Bacmeister and Maximiliano Doremberg (González Pacheco 1983; de Vos 1988a). These men, like many others situated in Mexico City or Northern Mexico, made enormous profits from the land in the Selva Lacandona without ever stepping foot in it (de Vos 1988a). Foreigners also acquired title to the lands of the Selva Lacandona. A member of a wealthy Spanish industrial family named Claudio López Bru, Marqués de Comillas, obtained a parcel of land measuring over 138,000 hectares in exchange for a land claim that he held in the state of Guerrero. The area located between the Río Lacantún and the Guatemala border in the south-

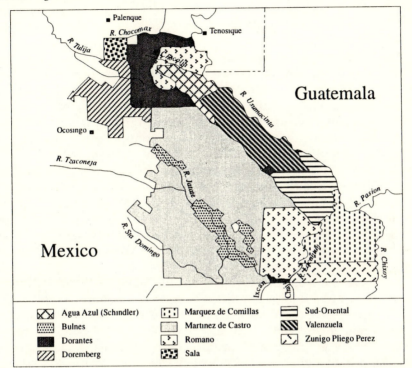

FIGURE 4.3 Surveyed properties in the Selva Lacandona in 1902 (Adapted from de Vos, 1992).

eastern corner of the Selva Lacandona continues to be referred to as the Marqués de Comillas region, despite the fact that the land became national property in 1955 (González Pacheco 1983; de Vos 1988a).

The purpose of acquiring these lands was not to obtain access to the resources, as it was for Bulnes, Valenzuela, and Romano, but rather to divide the property and sell off parcels at a profit to plantation owners, ranchers, and logging companies (de Vos 1988a). Although the contracts sometimes stipulated that the lands should be colonized, that was never the intention of the buyers. Instead, much of the land, particularly in the northern municipios of Palenque and Chilon, was sold off to ranchers or to North American companies interested in establishing rubber plantations. Colonization corporations such as the *Compañía Agrícola y Colonizadora de Tabasco y Chiapas, S.A.* were established and used to facilitate transactions with foreign capitalists searching for land investments (González Pacheco 1983). The allocation of surveyed land among speculators and logging companies in 1902 is illustrated in Figure 4.3. By this date, the entire Selva Lacandona had been divided among less than a dozen proprietors.

Timber Extraction

During the period between 1863 and 1914, referred to by González
Pacheco (1983:53, own translation) as "the golden era of the mahogany
trade," the Selva Lacandona became economically one of the most
important regions of southeastern Mexico. The profitability of the
logging industry during these years can in large part be attributed to
low capital requirements coupled with free transportation of resources
along the extensive river system of the Usumacinta basin and the
exploitation of labor. By the end of this prosperous era, the abundant
supply of trees located near the rivers had been exhausted. For logging
to continue, less accessible areas had to be opened up, and this would
require heavy machinery and enormous investments in infrastructure.
Such investments would add substantial costs to production. Rather
than make these investments, many private companies abandoned
their ventures in the Selva Lacandona.

Although the tropical hardwoods of the Selva Lacandona were in
many ways a free resource to the logging companies, there remained a
fine line between profit and failure in the industry. In the earliest
years of timber exploitation, distinct practices were followed to ensure
profitability. The first step involved obtaining permission to cut from
the government authorities of Chiapas. Many of the tabasqueño logging
companies maintained an agent in San Cristóbal solely to take care of
this. Such permission was only granted to those who had title or
pending title to the land. The next step would be to establish a sales
contract with a foreign buyer. No wood would be cut prior to the
existence of this signed agreement, which stated the volume of wood,
the size and quality requirements, and the price. Forty percent of the
sales was advanced to the logging company, with the remainder
contingent upon delivery (de Vos 1988a).

Prior to signing the contract, the logging company would have
explored and inventoried the area to be logged, thus the company
already had a clear idea of what type and quality of wood could be
delivered. The inventory was carried out by a "mahogany hunter," a
person with experience in locating mahogany trees. The mahogany
hunter would work his way to the tallest tree in the area, and ascend it
to obtain an aerial view of the canopies. Inventories were generally
carried out in August, when the mahogany trees were shaded a pale
reddish color and thus could easily be spotted. The patch with the most
abundant mahoganies would be visually selected, then trees within the
patch would be marked and their yields estimated. The next step was
the most critical, for it involved locating the nearest stream or river so

that the logs could be transported out of the forest. This task could take some time, given the density of the forest. Finally, a team of machete-wielding trailblazers would cut a path through the forest so that the trunks could be dragged to the river (de Vos 1988a).

After the best stand of mahogany trees was located, a central camp would be set up close to the river. Once established, the central camps rarely changed location. These camps formed part of an administrative network for the logging company,[3] and the managers of logging operations were stationed at these central camps. According to de Vos (1988), the camp looked very much like a small village. Recently recruited laborers would arrive at the central camp and from there be distributed to different districts in the zone. Provisions and materials were transported by caravan to the central camp, where the company store and supply warehouse were situated. The pens for the oxen and other beasts of burden were located at the central camp, as were the office and house of the manager, the bungalows of the employees, and the sheds and galleys of the laborers.

Using the camps as a base, axemen would fell the selected trees, and the trunks would then be dragged by teams of oxen through the forest to the nearest water channel. The logs were collected at the side of the stream or river and held there until the start of the rainy season. With the onset of the rains, the streams and rivers would swell and allow the logs to float unobstructed to the Río Usumacinta, and from there to Boca del Cerro in Tabasco, where they would be retrieved. Calculating the exact location and timing of the release of the logs downstream was a critical step in the process of timber extraction from the Selva Lacandona. In some cases, the rains came late or came with a false start, such that the rivers would rise and then descend again rapidly, leaving the logs stuck on the sides to slowly rot. In extreme years, when rains arrived late and were insufficient to fill up the rivers, no trunks made the journey out of the Selva (de Vos 1988a).

A guide would usually travel in a canoe alongside the logs, ensuring that they made it past difficult parts of the river. Once the current became strong enough, the guide would leave the logs to float on their own. Prior to the journey down the river, logs would have been marked for later identification. For example, logs harvested from national lands would be marked "M.N.P." for *Madera Nacional Permiso*, while those taken from land owned by the logging company were marked with "M.P.P" for *Madera Propiedad Permiso*. A single letter following these marks would signify the logging company (Moscoso Pastrana 1966).

Each year, thousands of mahogany and cedar trunks successfully completed the journey out of the forest to Boca del Cerro, Tabasco.

There, local residents equipped with small canoes and large hooks would collect the logs and wait for company representatives to claim them and pay the one peso per trunk fee (González Pacheco 1983:143). Once a company had claimed its trunks, large rafts would be constructed by connecting 50 to 100 logs with vines or chains. These rafts would be towed by steamer or floated to their final destination prior to export (de Vos 1988a). The final destinations for the tropical hardwoods of the Selva Lacandona included the ports of Frontera, Laguna de Tér-minos, and Ciudad del Carmen, all of which are located in Tabasco.

When the rafts arrived at the port, a company representative would meet them and the logs would be dismantled and cleaned up in preparation for quality evaluations. The trunks that were not split, rotted, or damaged along the journey would be measured and classified based on size. According to experienced estimates, only half of the trunks cut in the forests of Mexico and Central America arrived at their destination (de Vos 1988a). This was the result of cumulative losses, from the cutting stage where unhealthy trees would be left to rot or trees that were damaged in the fall would be abandoned, through the transportation stage where numerous losses or damages occurred, and finally to the measurement and evaluation stage where many trunks simply did not meet export requirements.

When all of the valuable trees surrounding the camp had been cut down, new districts would be opened up and smaller camps would be established. The central camp would then be converted into the administrative offices for the entire zone, and the secondary camps would become the functional logging camps. The functional and admin-istrative setup of timber operations resulted in a large number of camps throughout the Selva Lacandona. Some of these later became villages or towns, such as the Bulnes camp in San Quintín.

In some cases, laborers who came to the area to work for a logging company remained after the operations had ceased. This was, how-ever, the exception rather than the rule. Given the choice, the majority of the loggers returned to their homes. In most cases, the labor that fulfilled the needs of the industry was neither voluntarily recruited nor justly compensated for the work.[4] The labor needs associated with timber extraction were enormous, since it relied primarily on human labor supplemented by animal traction. Consequently, unethical means were used to conscript workers. According to Benjamin (1981), the logging industry serves as an excellent example of what can become of capitalist production when it is out of the view of the public and is tolerated by the political structures.

Although the earliest workers in the logging camp came from Tabasco, by 1910 most of the labor supply was contracted from indigenous workers in Ocosingo (Moscoso Pastrana 1966). Representatives or contractors from the logging companies, known as *enganchadores*[5] would supply money or alcohol to the Indians during local fiestas, so that they would be indebted and therefore had to sign contracts for one or two years of work (Benjamin 1981). The enganchadores also bought workers directly by paying the fines of Indians serving time in jail. In addition, President Porfirio Díaz ordered rebellious Maya Indians from Yucatán and Yaquis from Sonora to serve in the logging camps (Benjamin 1981).

Once working in the logging camps, laborers quickly became tied to their employers through a system of debt peonage. Forced to purchase goods in the company store at elevated prices that their low wages could never cover, workers became increasingly indebted, and it subsequently became more and more difficult for them to leave the camps. Indebted servitude as a form of labor control was quite common in Chiapas during the 19th century (Benjamin 1989). In fact, the practice earned Chiapas the reputation of being Mexico's slave state.[6]

The labor practices of the logging industry experienced a blow in 1913, during the Mexican Revolution. General Luis Felipe Domínguez, a rancher from Tabasco, entered Chiapas and journeyed from logging camp to logging camp, liberating workers along the way. This "Usumacinta Brigade" decreed the absolute liberty of the workers, the abolition of all workers' debts, and the execution of the administrators and overseers. The itinerant rebellion put the logging camps out of business for only two or three years, but in doing so it became a legend in Indian Chiapas (Benjamin 1989).

Exploitation of the workers was facilitated through mechanisms developed by landowners and politicians to create a cheap and available labor force. However, the conditions of loggers improved as a result of several campesino-based social movements that emerged in Chiapas after the Mexican Revolution. Through the 1914 *Ley de Obreros*, company stores were abolished and indebted servitude was eliminated. Working conditions and wages also improved over time, although they were still far below what might be considered fair (González Pacheco 1983). Nevertheless, by the time that these improvements were introduced, the logging industry had already fallen into decline, and labor requirements had substantially diminished.

The End of an Era

According to logging practices in the early part of this century, the first trees to be cut down from the forest were the largest ones located closest to the rivers. Those trees earned the greatest profits relative to their extraction costs. Over time, however, the desirable trees were found at increasing distances from the rivers, and logging companies were forced to expend more and more resources to extract them. As long as there was an external market, logging remained profitable. However, when the European market dried up during World War I, the logging industry fell into a slump and never regained its earlier strength. Although the North American market replaced Europe as the main destination for timber extracted from the Selva Lacandona, it became increasingly unprofitable to export timber.[7]

Although some of the original logging companies stayed in business for several decades longer, many of them sold off their land and abandoned their camps in the Selva Lacandona. New and smaller companies emerged, but they would never see the profits that existed during the best years of the mahogany trade. These companies were forced to venture into the more difficult-to-access areas — the ones that had been avoided for years in earlier timber operations. In order to modernize production methods (for instance, by replacing oxen with heavy machinery), enormous investments in infrastructure would be required (González Pacheco 1983). Aside from the expense of modernization, the practicality of it was limited by the remote location of the industry within the Selva Lacandona.[8]

Logging companies slowly ceased to operate in the Selva Lacandona during the first half of the twentieth century. By 1949, the year in which the Mexican government passed a law prohibiting the export of unprocessed timber, the few remaining logging companies had already folded up operations and left the forest. The golden era of the mahogany trade was clearly over.

According to de Vos (1988), the Selva Lacandona returned to its virgin state between 1949 and 1954, and was inhabited exclusively by about 400 Lacandón Indians. This is probably an exaggeration, considering that some of the employees of logging companies remained in the forest after the camps closed down. For example, Moscoso Pastrana (1966) mentions that after the logging camps began to close in 1944, many workers settled along the Lacantún river and the upper Usumacinta. The one-time loggers began to cultivate corn, beans, and sugar cane in places known today as Pico de Oro, El Cambio, Agua Azul, Macachive, and El Desempeño. In addition, colonization of the forest periphery had already begun in the 1940s, with the first settlers arriving from the Highlands of Chiapas. Although the forest in 1950 was not much more accessible than it had been one hundred years earlier, the first signs of settlement were already apparent.

Between 1949 and 1964, land titles in the Selva Lacandona underwent numerous transfers. González Pacheco (1983) documents how one North American enterprise, the Vancouver Plywood Company, obtained titles to a large part of the forest with the intention of massively exploiting its timber resources. A Mexican front company known as Maderera Maya was established to carry out operations that included harvesting both hardwoods and secondary species, the latter of which would be used in the production of wood pulp. The envisioned venture never took off because the federal government denied the company permission to construct a required timber processing plant. According to Gómez-Pompa (1992), the request was refused in response to a media campaign accusing the company of planning to destroy Mexico's forest resources. This early outcry against deforestation led to an abandonment of plans for large-scale timber exploitation.

A Contentious Period

Logging of the Selva Lacandona got underway again in 1964, this time through the Weiss Fricker Mahogany Company of Pensacola, Florida. The company, which had extensive experience in logging tropical hardwoods throughout Central America, had contracted logging rights to the Selva Lacandona from the Vancouver Plywood Company and its Mexican affiliate, Maderera Maya (González Pacheco 1983). It set up another Mexican front company, Aserraderos Bonampak, to carry out the actual logging. In contrast to previous logging operations, Aserraderos Bonampak used modern production techniques to gain access to untouched areas. A network of roads were built into the most mahogany-rich areas of the forest, including a road from Palenque to the company headquarters in Chancalá. Such roads paved the way for colonization, and led to the settlement of ejidos such as Monte Líbano, El Diamante, El Prado, and El Retiro, along with many others (González Pacheco 1983).

The areas first exploited by Aserraderos Bonampak were rich in resources, as they had never been logged before. However, as time passed the valuable timber could only be found deeper and deeper within the forest, and this meant increasingly expensive investments in road construction and maintenance. The cost of further expansion outweighed the financial benefits, and after some deliberation, in 1972 Weiss Fricker decided to sell Aserraderos Bonampak to the Mexican-owned *Nacional Financiera, S.A.* (NAFINSA). In 1974, a presidential decree created the *Compañia Forestal de la Lacandona, S.A.* (COFOLASA) and a complementary wood-processing company, *Compañia Triplay de Palenque, S.A.*, both under the control of the NAFINSA. These companies were created to replace private timber companies with state-controlled enterprises. Ostensibly, the main objective was a rational utilization of forest resources that would contribute to an improved standard of living for those in possession of the resource (Carmona Lara 1988). The move was also part of the Echeverría government's national forest exploitation policy "that saw the creation of many enormous forest parastatals controlled by rent-seeking bureaucrats" (Bray 1997:4) .

COFOLASA paid communities for 25 years of logging rights, then contracted out the actual logging in return for credits and a part of the profits (Godinez Herrera 1989). However, part of the royalties were paid to the *Secretaría de la Reforma Agraria* (Ministry of Agrarian Reform), which would later disperse the money in the form of community services. Using the prevailing silvicultural techniques, COFOLASA haphazardly laid out logging roads and then logged the best trees, leaving behind the diseased and genetically poor trees (Bray 1997).

Politics has always been at the heart of land transactions in the Selva Lacandona, and of forestry policy as well. The exploitation of forest resources in Chiapas has served as an important element in its economic integration with the rest of Mexico. As such, the loss of valuable forest resources concurrent with colonization of the forest in the 1960s did not escape government notice. During the same period, the Lacandón Indians were lobbying to secure title to the land that they still controlled in the Selva Lacandona. Much of this lobbying was done by Gertrude Duby Blom, a Swiss-born resident of San Cristóbal de las Casas. Along with her husband, Franz Blom, she had been one of the earliest to criticize deforestation of the Selva Lacandona and bring the plight of the Lacandón Indians to international attention (Woodward and Woodward 1985).

In response to the rapid loss of forest resources due to colonization, and to placate the demand for secure land titles by the Lacandones, a 1972 presidential decree gave legal title to 614,321 hectares of forest to 66 Lacandón heads of household (Diario Oficial 1972).[9] The ostensible basis for this decree was that the group had been in possession of the land "since time immemorial."[10] Thus, despite the fact that portions of the land titled to the *Comunidad Lacandona* were already populated with migrants, the government transformed the Lacandones into the largest latifundistas of the region.

This massive transfer of property rights had important political implications. First, it meant that approximately 6,000 Tzeltal and Chol colonizers in 23 ejidos were now illegal squatters in the Selva Lacandona, despite the fact that several communities had already been granted ejido status by the government. To address this situation, in 1975 the Mexican government relocated 13 ejidos to the planned community of Palestina, also known as Nuevo Centro de Población Velasco Suárez. The remaining 10 ejidos were resettled at Frontera Corozal, sometimes called Centro de Población Luís Echeverría. Palestina was located on a 25,000 hectare site to the west of Yaxchilán, whereas Frontera Corozal was established on 20,000 hectares located on an isolated, swampy plain along the Río Usumacinta (Price and Hall 1983).

The outcome of this resettlement scheme was disappointing for many of those involved, and it created a great deal of resentment toward the government as well as the Lacandón Indians. A number of conflicts erupted between the Lacandones and groups claiming title to the same land (see Burguete Cal y Mayor 1978). Some of the migrants who had refused the government's resettlement offer invaded national lands within the Comunidad Lacandona, and others petitioned the government to legitimize their land claims (Union de Uniones Ejidales y Grupos Campesinos Solidarios de Chiapas 1983). Although the Mexican army reportedly burned illegal settlements during the late 1970s, migrants were apparently allowed to return after COFOLASA finished logging in certain sections (Price and Hall 1983).

Indeed, it appears that the Comunidad Lacandona was conceived as a means for securing state control over forest resources in the Selva Lacandona. For the Lacandón Indians, land rights to a large area of the Selva Lacandona meant that they were free to sign lumbering contracts with COFOLASA, the state-run lumber company. According to Nations (1979:112), "[i]t seems likely that government officials were more eager to deal with 350 illiterate Lacandones than with 50,000 Tzeltal and Chol immigrants; sure enough, these contracts opened the zona Lacandona to the exploitation of the vast stands of mahogany and tropical cedar that had escaped earlier lumbering operations."

In return for rights to log the forest for 25 years, the Lacandón Indians received gifts of clothes and medicines, as well as promises of cash payments. However, not all of the timber royalties were given directly to the Lacandones: 70 percent were placed in a common fund controlled by NAFINSA and administered by the *Fondo Nacional Para el Fomento Ejidal* (FONAFE), and the remaining 30 percent were given to the Lacandones in biannual payments. The royalties controlled by NAFINSA were used by a number of government agencies with the intention of helping the Lacandones "improve their lives" (Nations 1979). A study by Godinez Herrera (1989) describes how paternalistic government agencies excluded the communities within the Selva Lacandona from active participation in the forestry sector, foreclosing the possibility of obtaining higher prices for their resources.

Nations (1979) describes the impacts of the sudden windfall on Lacandón subsistence practices as disastrous. Not only did the Lacandones become reliant on purchased foods rather than their own harvests, but the new condensed settlement patterns proved to be breeding grounds for infectious diseases such as influenza, intestinal parasites, measles, salmonella, and shigellosis. As a result, the Lacandones spent a good portion of their royalties on medical treatments. They also used their new income to purchase consumer goods such as trucks, watches, and jeans. With the exception of a few families who continued to maintain a traditional lifestyle outside of the settlements, the generous land grant perhaps unwittingly brought the Lacandones into the twentieth century, and opened up a vast new supply of mahogany and tropical cedar trees to exploitation.

Although COFOLASA received logging rights from the Lacandones, it still had to confront the previously settled colonizers regarding access to resources. In particular, COFOLASA had to overcome resistance from ethnic groups settled within the Comunidad Lacandona who were dissatisfied with the existing distribution of royalties. The protests involved physical blockades of logging roads, which meant a reduction in the number of logs extracted and a decrease in revenues. The conflict was settled in 1977, when Chol and Tzeltal colonizers were incorporated into the Comunidad Lacandona and granted a portion of the royalties (Burguete Cal y Mayor 1978). In the meantime, continuing colonization of the Selva Lacandona resulted in further timber loss, since most migrants simply burned the forest to make way for their milpas. A consequence of this was that forests in Chiapas were disappearing at an unprecedented rate. State-wide losses were estimated to be from 50 to 66 thousand hectares a year between 1976 and 1992 (Cruz Coutiño and Parra Chávez 1994).

The widespread forest loss was not matched by a corresponding gain in the forestry sector of Chiapas. The forestry infrastructure was, in fact, highly underutilized. Sawmills were operating at 51.4 percent below capacity, whereas plywood and board factories were running as much as 90.4 percent below capacity (Cruz Coutiño and Parra Chávez 1994:24). COFOLASA operated at a loss during most years, and was unable to fulfill its supply commitments to Triplay de Palenque. In 1980, the company was acquired from NAFINSA by the state of Chiapas, which held it under the *Corporación de Fomento de Chiapas, S.A. de C.V.* (CORFO) until it ceased operations in 1989. COFOLASA was formally dissolved in 1991.

In the final years of logging, the number of mahogany and tropical cedar trees harvested by COFOLASA declined drastically due to their increasing scarcity and implementation of conservation measures. In contrast, the harvest of common species increased as new uses for them were developed (Table 4.1) (Carmona Lara 1988:138). Labor unrest also affected productivity, and much wood was left to rot during prolonged strikes (Rojas 1995).

While a profitable logging industry could not be adequately revived in the Selva Lacandona, COFOLASA left its mark on the region through the construction of roads. Accurate data on the number and quality of roads constructed by COFOLASA are not available. However, by most accounts the company opened up extensive parts of the forest by constructing roads into once inaccessible areas. It was also responsible for building roads to the resettlement areas of Palestina and Corozal (Burguete Cal y Mayor 1978).

The decreasing profitability of logging within the Selva Lacandona was largely a result of increasing production costs associated with access and transport, combined with a decline in the harvest of profitable hardwoods and general mismanagement.[11] The situation paralleled a general malaise in the forestry sector of Mexico, which

TABLE 4.1   Extraction of timber by COFOLASA during the 1980s.

| | Timber Sold | | | Royalties to |
Year	Precious (Board-feet)	Common (Board-feet)	Employment (persons)	Comunidad Lacandona (old pesos)
1983	2,318,850	120,711	363	56,692,000
1984	2,331,353	256,556	430	116,981,000
1985	1,221,552	684,885	506	159,450,000
1986	1,037,369	383,585	415	65,000,000
1987	902,486	1,365,382	416	122,000,000
1988	332,222	1,173,499	273	160,000,000

Source: Carmona Lara, 1988.

had come to rely heavily on government subsidies (Godinez Herrera 1989). In general terms, Mexico's forestry sector was underdeveloped, and excluded local communities from the activities of the sector (Halhead 1984; 1992).

In 1989, governor Patrocinio González of Chiapas announced a new forestry policy which included a complete ban on logging, or *veda forestal*. Laws were changed, and any person guilty of a crime or offense relating to the new forestry laws would be subject to prosecution. This reorientation was in large part due to uneconomical and wasteful practices followed during the 1980s by both private companies and by the state itself (Cruz Coutiño and Parra Chávez 1994). However, the logging ban could also be seen as a response to the growing national and international concern over the rapid deforestation of the Selva Lacandona. Supporting the environmentalist image of President Carlos Salinas de Gortari, González Garrido's timber ban could be considered a rather painless concession to environmentalists, as logging had ceased to be profitable to the state anyway.

Although a timber ban was declared, it was never legally enacted, and forest policy remained somewhat ambiguous. According to Bray (1997), under Mexican law a ban requires a series of time-consuming studies to justify it, which González Garrido did not bother to carry out. Nevertheless, a series of measures were passed in accordance with an agreement signed on May 18, 1989 by the Mexican federal and Chiapas state governments, creating the *Coordinación Forestal del Estado*, whose principal function was to distribute (or withhold) forest exploitation permits and formulate programs for forest exploitation and conservation (Bray 1997). The measures touched on most aspects of forest exploitation. For example, in 1990, the government of Chiapas established an obligatory registration law for chainsaws, with the aim of controlling ownership, use, commercialization, and reparations (Cruz Coutiño and Parra Chávez 1994).

By 1993, 1,838 chainsaws were registered in Chiapas. Of the 49 sawmills operating in 1989, all but four had closed by 1993. In addition, all of the state's four plywood factories were shut down (Bray 1997). The logging industry of Chiapas, in a crisis prior to 1989, thus died quickly. More painful was the impact of the ban on local communities, as it meant that they were prohibited not only from removing timber, but also from clearing new land for cultivation.

Denying local groups the right to remove trees proved to be a source of conflict, resulting in a number of confrontations between locals and government officials. The most frequently cited example involved the community of Nuevo Chihuahua in the Marqués de Comillas region. In

July, 1991, state police responded to reports that campesinos from the community had exceeded their authorized logging limits, and had logged live trees as well. The police attempted to confiscate the timber, and a clash resulted, with the community detaining 63 people in the *Casa Ejidal* and demanding an audience with the governor to appeal the confiscation. They argued that the timber had beeen cut with permits granted prior to the veda forestal, and that COFOLASA owed them money for the wood (Bray 1997; Harvey 1997). An estimated 393 protesters banded together as the Movimiento Campesino Regional Independiente (MOCRI) and traveled to Mexico City to demand that President Salinas address the issue of marketing already-dead wood from ejidos. Rather than securing a solution to their problem, they instead suffered human rights abuses in Palenque (Fernández Ortiz et al. 1994; Harvey 1997).

The forestry ban was reversed in 1994, when the government granted permits to 18 ejidos in the Marqués de Comillas region to remove trees that had fallen or died, and sell the wood. This decision, taken by the *Secretaría de Recursos Hidráulicos* (SARH), was in part a response to the Zapatista Uprising and the need to appease non-Zapatistas in the Selva Lacandona. Beneficiaries of the permits included both supporters of MOCRI and the PRI-affiliated *Union de Ejidos Juan Sabines*. According to Harvey (1997), there was little regulation or control of how the ejidos used the new permits, and some ejido officials illegally sold both permits and signatures to non-ejidatarios and outside entrepreneurs.

Over the past three years, significant amounts of forest have been logged both legally and illegally in the Marqués de Comillas region. Harvey (1997) argues that logging has proven to be the most economically viable option for many of the campesinos in the Marqués de Comillas region, in spite of the illegality of logging live trees. Although only 22,899m^3 of dead mahogany trees were authorized for exploitation, a 1995 inventory conducted by SEMARNAP[12] found that 120,439 m^3 of *live* mahogany trees were actually cut (Harvey 1997). Most ejidos surpassed their allowable cut and instead absorbed the fines imposed by government. They were able to do this because the principal buyer of timber, a Palenque-based corporation called CARPI-CENTRO, simply reimbursed the campesinos for the fines by paying a slightly higher price for the timber (Harvey 1997).

The government has recently responded to the growing conflicts over logging by establishing a *Plan Piloto Forestal* in the Marqués de Comillas region. According to this plan, modeled after a succesful program in the Quintana Roo, ejidos would manage their forests for their own benefits, at the same time conserving the forest and its associated biodiversity. Participants will eventually be encouraged to process the timber, adding value and generating profits. Twenty-six species of tropical woods, including cedar and mahogany, are eligible for exploitation under the Plan Piloto Forestal. Although a number of ejidatarios are prepared to participate in the Plan Piloto, as of June 1997 the project was still awaiting the final approval by the environmental ministry, SEMARNAP (Neubauer 1997).

The reversal of the forestry ban was heavily criticized by environmentalists, and it has led to tensions between supporters of conservation and proponents of rural development, both within and outside of the government. Whether these tensions are alleviated through the successful management of forests under the Plan Piloto Forestal remains to be seen. Nevertheless, the existing tensions emphasize the dichotomy between environmental and social struggles in the Selva Lacandona, which are explored in greater detail in Chapter 8.

Oil Exploration

Petroleum is another resource which must be considered a part of the more recent history of the Selva Lacandona. Like timber extraction, the activities of the Mexican state oil company, *Petróleos Mexicanos* (PEMEX), have been widely held responsible for deforestation in the Selva Lacandona. While the environmental record of PEMEX in other areas of southeastern Mexico has been justly criticized (Toledo et al., no date), its operations in the Selva Lacandona were considerably less destructive, largely because they were limited to exploratory activities. In fact, a study by Hernández Escobar (1992) concluded that aside from some local pollution and illegal trafficking in endangered species, the environmental impacts of PEMEX activities were insignificant. The direct impacts of petroleum exploration on deforestation were confined to the areas where the wells were drilled, and to the networks of roads constructed to access the wells. In fact, the most lasting effect of petroleum exploration in the Selva Lacandona can be considered the increased number of roads in the region. These roads facilitated commercial activity in existing ejidos and communities, and opened the area up to further colonization.

Oil was discovered in the Selva Lacandona during the 1970s, after reserves were found in the Petén region of Guatemala. As this discovery coincided with a period of high oil prices, the Mexican government resolved to use the country's oil reserves as a catalyst for economic growth. During the *sexenio*[13] of President Lopez Portillo (1976-1982), the production goals of PEMEX included the following:

- an increase in the production of crude oil from 700,000 to 2.2 million barrels per day by 1982;
- an increase in gas production to 4,000 million cubic feet per day;
- an increase in exports to 1.1 million barrels per day;
- a doubling of the capacity of refineries;
- a tripling of the capacity of the petrochemical industry;
- an extension of the transportation network;
- an increase in oil exploration programs (Hernández Escobar 1992:28).

The existence of potentially large reserves in the Selva Lacandona meant that this region could be integrated into Mexico's development plans. Southern Mexico already provided a major source of oil revenue for Mexico, both from off-shore and terrestrial reserves. Since 1972, PEMEX has exploited oil resources in Chiapas, Tabasco, and in the Bay of Campeche. In Chiapas, most of the petroleum activity is concentrated in the northern part of the state, in an area known as Reforma. It forms part of the Cactus production district, where hydrocarbons are recovered from the gas produced in association with crude oil extraction. In fact, Chiapas has been Mexico's largest producer of gas, providing on average 14 percent of the national supply (Hernández Escobar 1992).

Petroleum reserves were discovered in several areas within the Selva Lacandona, and camps were subsequently set up to explore the potential supply. PEMEX established an administrative headquarters in the town of Ocosingo, along with Campo Nazareth near Altamirano, Campo San Fernando (2,723 km²) in the Comunidad Lacandona, and Campo Lacantún (2,250 km²) in the Marqués de Comillas region. Most of the public attention has focused on the last area, where PEMEX has been the most active.

For PEMEX, the existence of sufficient infrastructure was a prerequisite to petroleum exploration and extraction. In the Selva Lacandona, this meant roads had to be cut through the forest. Campo Lacantún was located 190 km southeast of the town of Palenque, and the existing road extended only as far as Frontera Corozal, which was about 80 km short of the camp. Furthermore, this road was in very poor condition. PEMEX, the *Secretaría de Comunicaciones y Transportes*, and the state government collaborated to lengthen and pave the road, and to construct a bridge across the Río Lacantún that would enable materials and equipment to be transported to the soon-to-be-established facilities. The bridge, finished in 1984, not only served to connect Campo Lacantún with Palenque, but it also facilitated the arrival of new groups of colonizers (González Ponciano 1990).

PEMEX established its center of operations approximately 8 km south of the ejido of Benemérito de las Americas, and began drilling two exploratory wells in 1984. Both wells produced oil. These promising results led to a second stage of exploration. Beginning in 1989, five new wells were drilled in the Marqués de Comillas region. There were signs that the Selva Lacandona might indeed contribute to Mexico's oil development.

Throughout the exploration process, PEMEX was seemingly sensitive to environmental criticisms, and it made numerous attempts to incorporate ecological considerations into its policies and management plans. Mexico's environmental ministry at the time, SEDUE, prohibited PEMEX from drilling exploratory wells in the nucleus of the Montes Azules Biosphere Reserve, and made the company agree to no hunting or capture of wildlife or vegetation by workers or contractors.

In 1984, PEMEX agreed to cooperate with state agencies in developing institutional mechanisms to supervise both the development and conservation of the Selva Lacandona (PEMEX 1984). In effect, PEMEX was committing itself to the co-management of social and ecological aspects of all regions that showed potential for petroleum production. What PEMEX found through its experience in the last frontier of the Selva Lacandona was that, regardless of the direct effects of its activities, local populations were using its presence as an excuse to solicit help to a myriad of local problems (PEMEX 1986). For example, PEMEX was expected to assist communities in the construction of roads and schools, and provide electricity and medical facilities (Hernández Escobar 1992).

The amount of deforestation created by PEMEX during its tenure in the Selva Lacandona was largely tied to the construction of roads, and to a lesser extent the clearings for wells and other constructions. For the seven exploratory wells established in the Marqués de Comillas region, a total of 112.6 hectares were cleared for roads, and 17.02 hectares for the wells. The total deforestation involved removal of an estimated 20,000 to 47,000 trees, corresponding to 0.06 percent of all of the Marqués de Comillas region, and 0.07 percent of the forested area (Carmona Lara 1988).

The roads constructed by PEMEX had far-reaching consequences. Previously isolated communities were integrated into the rest of

Chiapas and Tabasco through the growing transportation network. The road to Palenque served as an artery for the transport of goods and people into and out of the region — a service that had previously been filled by boats, beasts of burden, or occasionally by charter plane. A hospital was constructed in Benemérito, providing emergency health service to communities throughout the Marqués de Comillas region.

PEMEX reportedly suspended or terminated its operations in the Selva Lacandona region in 1993, for reasons that remain somewhat ambiguous. According to some sources, PEMEX experienced problems with the Zapatista guerrillas in late 1993. Whether the withdrawal applied to the entire region or just the Marqués de Comillas area is also uncertain. However, regardless of the presence or absence of PEMEX administrative offices and personnel, the roads that were constructed to facilitate oil exploration remain intact, and the oil resources underlying the Selva Lacandona have been integrated into calculations of Mexico's energy resources. Whether Mexico's long-term energy strategy includes the development of these resources remains to be seen.

The history of logging in the Selva Lacandona and the activities of PEMEX in the region have been described in order to illustrate the role of different production processes and their uneven contributions to deforestation. Early logging was highly selective and concentrated along the rivers of the Usumacinta basin. Despite its immense profitability and notoriety for labor abuses, it was not responsible for the massive deforestation that is often attributed to it. Later logging efforts, on the other hand, were far more destructive. Moreover, these operations, along with oil exploration, paved the way for colonization, a process which would eventually compete with logging for the resources of the Selva Lacandona. More recently, the role of local communities in logging has become the focus of concern. After more than one hundred years of control by private companies and the Mexican and Chiapas government, communities are demanding a more active role in the control and exploitation of forest resources. This community assertiveness comes at the same time that the environmental community has garnered international support for protecting the Selva Lacandona.

Roads to Destruction

The preceding sections explained how extractive industries became established in the Selva Lacandona, eventually paving the way for large-scale colonization and deforestation. The role of roads in the development and destruction of the Selva Lacandona is, however, a sensitive issue. From an environmental perspective, the existence of roads into a tropical forest undoubtedly contributes to its rapid demise. From a social perspective, roads provide access and mobility, enabling residents to transport crops and goods to markets, take advantage of medical services available in larger towns, or take care of official matters regarding land tenure, credits, and other business in Tuxtla Gutiérrez. While environmentalists may cringe at the sight of new road construction, from a development perspective it respresents progress and integration with the world outside of the Selva Lacandona.

With the exception of those constructed and used by extractive industries, the development of roads in the Selva Lacandona has been an unorganized and poorly planned endeavor (Figure 4.4). Roads promised by government officials were extremely slow in coming, contributing to disappointment and disillusionment with the government among local communities. In many cases, roads were evident on state maps years before they were actually constructed, providing an image of development that was far from the reality experienced by residents in the forest.

For settlers in Las Margaritas, a road extending from the Lagos de Montebello to Flor de Café was scheduled to be constructed in 1966. However, actual construction did not begin until 1970, and it was not completed until the early 1980s (Paz Salinas 1989). Likewise, a road from Ocosingo to San Quintín was started in 1979, but not completed until 1992 (Leyva Solano and Ascencio Franco 1996). The most rapid road-building activity was actually tied to the military threat along the Guatemalan border in the 1980s. In response to this threat, the government allocated nine million dollars to the construction of a paved road along the Guatemalan border (Dziedzic and Wager 1992). Along with the road, ten communities were settled along the border to establish a Mexican presence in the area (González Ponciano 1990).

To overcome government inertia, some communities pooled resources to construct roads from distant villages to a larger main road. For example, ejidatarios from San Pedro Yutniotic, located along the Rio Santo Domingo south of El Pacayal, spent thousands of dollars to construct a 6-kilometer road to connect their community to the main road. Resources to finance the road came in part from remittances sent by community members who had temporarily migrated to the United States for employment. Other communities remained dependent upon footpaths that were muddy and difficult to negotiate during much of the year. Unless an individual or community could afford to pay for passage on the small Cessna airplanes servicing the region from Comitan and Ocosingo, a trip to the municipal centers and markets of

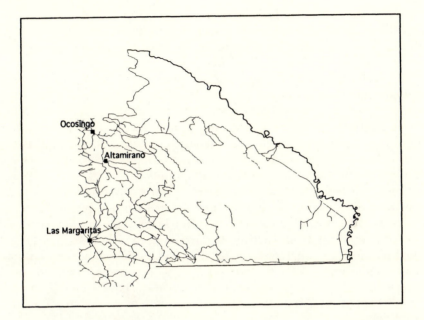

FIGURE 4.4 The road network in the Selva Lacandona region by 1992, up to Ocosingo (Source: Ecosur, 1997).

Las Margaritas, Ocosingo, Palenque, Altamirano, Comitán, and San Cristóbal was an arduous journey.

Roads have played a large role in differentiating communities in the Selva Lacandona. Those communities that are well connected to roads are generally better off. For example, the paving of the highway between San Cristóbal de las Casas and Palenque, which runs through Ocosingo, helped to integrate the communities living close to it with the state and nation, whereas the settlements that were distant from the road remained isolated (Leyva Solano and Ascencio Franco 1996).

The pace of road construction in the Selva Lacandona has accelerated dramatically since 1994. An increasing military presence in the region combined with renewed attention to the marginalized conditions of the people living in the Selva Lacandona have resulted in a flurry of construction activity. Roads have been paved and extended, in some cases at enormous costs. For example, the road from Comitán to Flor de Café has been extended to reach the once-isolated community of Peña Blanca (population of 153 in 1990), located atop a mountain ridge, and continuing on to the ejido of Ixcán (population of 417 in 1990). Paved roads have reduced travel times considerably, increased the amount of bus, pickup truck, and *combi* traffic, and enabled far greater access into and out of the region than existed even five years ago. The roads have also facilitated the movement and activities of the Mexican military, an issue discussed in Chapter 7.

Extractive industries have contributed to the network of roads into the forest, but they cannot be held responsible for all of the roads in the Selva Lacandona. Although roads facilitate colonization, it is difficult to conclude that colonization is a direct outcome of roads built by extractive industries. Such an assertion ignores the social and economic realities that truly drive land use changes. In many cases, roads were constructed as a result of colonization, rather than as a precedent to it. There are diverse explanations for the large-scale immigration of peasants coming from within Chiapas, from other states within Mexico, and from Guatemala. What the explanations have in common is that they can all be tied to the social, economic, political, and agrarian conditions existing outside of the Selva Lacandona region. In the next chapters, some of the factors driving land use changes will be examined.

5

Agricultural Changes and Continuities

The Agricultural Frontier

For most of the last century, the wealth of the Selva Lacandona was measured by the value of products that could be extracted and sold. However, Mexico's growing emphasis on agricultural production, in combination with a resistance among Chiapas landholders to agrarian redistributions, has steadily pushed the agricultural frontier into what is considered by some to be one of the most marginal agricultural areas in southeastern Mexico — the Selva Lacandona region of Chiapas. Colonization of the Selva Lacandona began slowly in the 1930s and intensified after 1960, when the region was opened up for agriculture in order to satisfy growing demands for agrarian reform. By the 1970s, the area had gained a reputation throughout Mexico as a region of economic opportunity, primarily due to the availability of land. As a result, thousands of migrants made the journey into the Selva Lacandona to establish permanent settlements. Concurrent with the expansion of colonization during the 1960s and 1970s, increasing amounts of forest and agricultural land were converted to pasture for livestock production. Over the past two decades, cattle ranching has become an important mode of land use in the Selva Lacandona, both on private ranches and ejidos. Like colonization, cattle ranching has resulted in deforestation of the Selva Lacandona.

Contrary to popular belief, deforestation in the Selva Lacandona has not been a haphazard, random process undertaken by growing populations of irrational peasants. Instead, it is the result of land use decisions based on economic realities. These realities are dictated in large part by agricultural transformations in Mexico and agrarian policies in Chiapas. In this chapter, important changes (and in some cases, lack of changes) will be described in order to provide a context for the dynamics of land use changes in the Selva Lacandona.

Agricultural Transformations in Mexico

The changes in land use that have occurred in the Selva Lacandona and Chiapas must be considered in the context of the transformation of Mexican agriculture as a whole. Over the past half century, this transformation has been one of the driving forces behind the colonization of Mexico's tropical forests. The transformation of Mexican agriculture has been described by Sanderson (1986:6) as the "product of a systemic internationalization of capital in agriculture and the long-term creation of a new global division of labor." Capitalist production has developed rapidly in Mexican agriculture since the Mexican Revolution (1910-1917). According to de Janvry (1981), this growth is based on a small minority of producers who operate large, modern, and efficient farms that produce a majority of Mexico's total agricultural output. Alongside these large-scale farms exist an overwhelming majority of producers who farm on small, often poor quality plots of land. Although these farms contribute to almost half of Mexico's corn production, they make relatively small contributions to national productivity, particularly in terms of export revenues.

At the turn of the twentieth century, Mexican agriculture was characterized by an extremely unequal distribution of land. Encouraged by government policies enacted during the Porfiriato, agricultural capitalism blossomed, benefitting the few large landowners who held a disproportionate amount of the land (Sanderson 1986; Cockcroft 1990). The majority of the rural peasants were landless and dependent upon the large estates for employment.

Peasants benefitted from the land reform laws passed after the Revolution, in theory if not in practice. For example, Article 27 of the Constitution gave the government the authority to expropriate parts of large landholdings and redistribute them to applicants who met certain qualifications (Yates 1981). In reality, however, "the revolutionary politico-bureaucratic elite actively promoted and participated in a form of agricultural development based on large private landholdings and capitalist accumulation. Land distribution, in the form of ejido grants, was undertaken only under pressure from organized groups of peasants or in response to de facto situations involving peasant possession of ancestral land" (Grindle 1986:62).

Land reform was not taken up in earnest until Lázaro Cárdenas assumed the national presidency during the 1930s. The agrarian policy adopted by the Cárdenas administration included:

- The ejido as mechanism for increased agricultural production;
- The organization and unification of the peasantry;
- The promotion of rural credit through the National Ejido Credit Bank;
- Preferred attention to land distribution and restitution;
- The expropriation of latifundio lands (Reyes Ramos 1992:60).

Between 1935 and 1940, almost 18 million hectares of land in Mexico were redistributed among 811,000 people as ejidos and communal lands (Hewitt de Alcántara 1973). In taking up the issue of land reform and institutionalizing it as part of the national program, the Mexican state created a consensus among campesinos and was able to legitimatize itself in the countryside (Grindle 1986; Reyes Ramos 1992). Equally important, these small farms accounted for over half of the value of farm output by 1940. Much of this was a result of the National Ejido Credit Bank, which made credit and technical assistance available to farmers.

After 1940, a counterreform movement emerged in Mexico regarding agrarian policy. This movement resulted in a slowdown of land distributions as well as the proliferation of legal rulings in favor of private landholdings. Policies were generally oriented toward promoting productivity on capitalist forms of landholdings rather than in the ejidal sector, which was considered to be unproductive and technologically backward (Grindle 1986; Reyes Ramos 1992). Such policies were facilitated by the emergence of the "Green Revolution," initiated with support of the Rockefeller Foundation and the Mexican government (Hewitt de Alcántara 1973). Its purpose was to employ modern science in the expansion of agricultural production, with the alleged goal of feeding the growing population. The research resulted in hybrid seeds, which if combined with water, fertilizers, and pesticides would guarantee a high-yielding harvest. At the same time, massive public investments were being made in the commercial farming sector, particularly in irrigation schemes, such that this sector was destined to benefit from the revolution. In fact, according to Hewitt de Alcántara (1973), the Green Revolution became synonymous with a capital-intensive path of development open only to those who controlled sufficient resources, including land and water, to make an investment in the relatively expensive inputs necessary for production.

Furthermore, the massive investments in large-scale agriculture were undertaken during a period when Mexico was undergoing rapid industrialization and urbanization, and experiencing a relative labor shortage and inflation (Hewitt de Alcántara 1973). Subsidized prices determined by the state discouraged the capitalist sector from producing basic crops for domestic consumption. Instead, production was directed toward export crops for the North American market. Ejidatarios and small-holders, unable to compete in the export market, were thus forced to cultivate crops whose prices were controlled in the interests of urban consumers (Grindle 1986). While increased export production through the modern agricultural sector was a key strategy in financing industrialization, it was the ejido and small-holder sector that made it possible to maintain low food prices, and hence low wages, for the labor force.

In the years since the introduction of high-yielding varieties and other Green Revolution technologies, agricultural output has expanded considerably. The productive agricultural sector kept food prices low, and enabled rapid industrialization to take place within Mexico (Cockcroft 1990). As a net exporter of food, Mexican agriculture was held as a model for the world to imitate. However, the remarkable growth evident between 1950 and 1970 slowed during the 1970s (Wellhausen 1976). Agricultural productivity was not able to keep up with an expanding population, and Mexico was eventually forced to import basic food crops, putting an end to the "Mexican miracle."

Decreases in corn production were at the heart of the agricultural crisis, even prior to the financial crisis that began in 1982. Corn is the mainstay of most rural households, and it holds cultural as well as nutritional significance. According to Hewitt de Alcántara (1994:18), the deepening "corn crisis" in peasant areas had a number of causes, including:

> . . . population growth in rural communities where resources are relatively fixed; the low profitability of corn production in comparison with other agricultural and livestock options and, over long periods, even in comparison with the level of remuneration for labor; restrictions imposed on the availability of family labor by the seasonal and permanent migration of able-bodied members of the household; and the expansion of urban areas, livestock operations, and cultivation of forage over large expanses which were previously dedicated to corn.

The impacts of the crisis were temporarily allayed between 1980 and 1982, when the government initiated the *Sistema Alimentario Mexicano* (SAM). This National Food System was a program aimed at achieving self-sufficiency in basic grains. Although short-lived due to the precipitous drop in oil prices in 1982, the increases in purchase prices for basic grains had the effect of lessening the traditional bias against small producers, and it stimulated the production of marketed foodgrains on rainfed lands (Barkin 1990). The program did not last long enough to reverse a basic trend, whereby corn production was relegated to a smaller proportion of the total available agricultural lands, much of which was of marginal quality (Hewitt de Alcántara 1994).

Agricultural production in Mexico has shifted increasingly to an export-oriented commercial agricultural sector, specializing in winter vegetables for U.S. markets. Meanwhile, the neoliberal path of development included the gradual withdrawal of state support for the cultivation and marketing of basic grains, which are the livelihood of the campesino sector. During the *sexenio* of Miguel de la Madrid, government subsidies for agriculture declined by an average of 13 percent per year, in striking contrast to increases of 12.5 percent per year during the 1970s (Harvey 1995). The price of inputs skyrocketed with the 1982 devaluation of the peso, while at the same time access to credits dried up for small-scale farmers and ejidatarios.

Neoliberal Reforms

Changes in agricultural policies accelerated under the administration of Carlos Salinas. These changes were in part dictated by the structural adjustment loans made by the international financial community. The World Bank played a particularly important role in reshaping Mexico's economic and agricultural policies, as it pressured the government to privatize state-owned enterprises and open markets to foreign imports. In short, it promoted a substantial restructuring of Mexico's policies:

> The World Bank conditioned a 1991 sectoral adjustment loan for Mexican agriculture on the implementation of a laundry list of specific measures, including eliminating import permits for more than a dozen food products, removing or slashing agricultural tariffs, canceling price controls on a range of basic food items, privatizing state-owned monopolies, and, most dramatically, eliminating price guarantees for corn (Barry 1995:44).

Reforms to Mexico's agricultural sector culminated in the passage of the North American Free Trade Agreement (NAFTA) and revisions to Article 27 of the Constitution, both of which were perceived by many as blows to the small-holder and ejidal farming sectors. These changes were considered particularly devastating to peasant farmers in the

Selva Lacandona, and have often been mentioned as an underlying cause of the 1994 Zapatista Uprising, which is discussed in Chapter 7.

NAFTA represented a far-reaching attempt to liberalize trade among Mexico, the United States, and Canada. Unlike many previous trade agreements, the agricultural sector was included in the negotiations, which took place between 1990 and 1993. This can be considered a radical departure from Mexico's earlier stance, which focused on self-sufficiency in basic grains as an issue of national security (Barkin and Suarez 1985). The NAFTA treaty includes a gradual phaseout of guaranteed prices for maize in Mexico. In place of guaranteed prices, the Salinas administration in 1993 announced the creation of the *Programa Nacional de Apoyos Directos al Campo* (PRO-CAMPO). This program is based on direct cash payments for seven crops, calculated on a per hectare basis (Harvey 1995:46). The payments are not earmarked for agriculture, and in fact have been used to purchase anything from food and clothes to televisions and satellite dishes. PROCAMPO can be considered another example of concessionary politics that attempt to divert attention away from the real implications of NAFTA.

The implications of the agreement have been summarized by Barry (1995:66), along with his prognosis for the future of Mexican agriculture:

> In broad terms, NAFTA resulted in the gradual opening of the Mexican grain market to U.S. exports in exchange for the opening of the U.S. fruit and vegetable market to Mexican exports. Also significant was the opening of the U.S. market for sugar and cotton and the Mexican market for dry beans, oilseeds, apples, meat, and dairy products. NAFTA's full impact on the U.S.-Mexico agricultural balance of trade will not be felt until 2009 when all tariffs are phased out. However, unless there is dramatic growth in Mexico's agroexport sector, it is likely that NAFTA; together with Mexico's own economic policy reforms, will weaken Mexico's farm sector, deepen its dependence on U.S. trade and investment, and widen its agricultural trade deficit with the United States.

Not all sectors within Mexico will be equally affected by NAFTA and the removal of guaranteed prices and subsidies. Small-scale farmers and semi-commercial producers, including most ejidatarios, may be hit particularly hard by an influx of cheap grains from the U.S. and Canada.[1] Maize is the mainstay of rural production systems in Mexico, and a small surplus is often what provides households with a cash income to purchase other foods, clothing, or medicine (Hewitt de Alcántara 1994). Low-priced imports will render this surplus production non-competitive. Yet these farmers will not be able to simply switch to alternate crops, such as fruits and vegetables for the export market. Such a switch assumes access to credits for seeds, fertilizers, and other inputs, as well as storage, transportation, and marketing facilities. Another solution considered by proponents of free trade involves a shift to wage labor. However, this option depends on the existence of reliable markets for labor (as well as guarantees against labor abuses). In remote areas such as the Selva Lacandona, producers are distant both from markets and sources of steady wage labor. In fact, most colonizers migrated to the forest in search of land in order to *escape* situations of low-paid wage labor.

The possibility of acquiring a piece of land to farm has been the strongest legacy of the Mexican Revolution for the rural population. Although agrarian reforms have been subject to the inclinations of post-Revolutionary governments, the commitment to the peasant sector has been a revered tenet of Mexican politics. In 1991, President Salinas pushed through reforms to Article 27 of the constitution, which basically ended the government's commitment to land distributions. The reforms gave ejidatarios the right to sell, rent, sharecrop, or mortgage their plots of land, and they opened up the possibility for joint ventures and contractual arrangements with private investors and stockholding companies, including foreign-owned. Ejidatarios also gained the right to collectively dissolve the ejido and distribute the property among members. Finally, the requirement that ejidatarios work their land to retain control was abolished (Barry 1995; Harvey 1995).

According to de Janvry et al. (1997), the reforms were initiated in response to the very decay of the ejido system within the global context of political and economic liberalization. Although modernizing the agricultural sector and attracting new investments in the countryside were regarded as necessary steps in the reform, they raised a number of concerns regarding the future of Mexico's rural population. These concerns, discussed by Barry (1995) and Harvey (1995), include the fear of a reconcentration of land, an increased probability of property foreclosures and loss of titles, the potential abandonment of pending land petitions, an increase in urban-rural migration, and an increasingly polarized countryside.

Although the actual impacts of the reforms have not yet become apparent, there is already evidence that they have triggered some important adjustments in farming strategies and have fostered increasing social differentiation. The reforms have also created serious difficulties for a large number of ejidatarios whose livelihoods and continued access to land are threatened (de Janvry et al. 1997). The reforms have contributed to a widespread perception that there is little space for small-scale or subsistence farming in Mexico's agricultural future.

The changes in Mexican agriculture over the past century can be summarized as increases in both extent and intensity of production, as well as a dramatic shift in the crop mixture toward commercial and export crops (Barkin 1990). The expansion of capitalist agriculture in Mexico has led to an increased focus on foreign markets, rather than on domestic consumption. The internationalization of capital that has accompanied this transformation has led to what Sanderson (1986) described as the new global division of labor, and what de Janvry (1981) called a disarticulated economy, whereby labor is producing goods for distant markets, creating a situation where there is no incentive to stimulate local consumption through wage increases.

Economic liberalization, along with the integration of Mexico into the new international division of labor, has created a system that is incapable of reproducing conditions for the survival of the rural population, and one that is fundamentally threatening to peasant agriculture and nutrition (Sanderson 1986:8). Barkin (1990) identified three effects of the trend toward capital-intensive exploitation: (1) the replacement of permanent agricultural jobs with machinery; (2) the substitution of employees and migratory labor for independent workers; and (3) a reduction in the ability of agriculture to absorb new workers.

The responses to these changes varies across social groups, individuals, and geographical areas. De Janvry et al. (1997) observed the emergence of a small, relatively successful entrepreneurial sector, and found the defining characteristics to be quite varied. They did note that access to credit, irrigation, and education were key elements for pursuing modernization and crop diversification. For the majority of campesinos who have been unable to adapt to the modern, capital-intensive agricultural sector, recent years have been marked by increased poverty and marginalization.

De Janvry et al. (1997) also found that different regions display different characteristics in terms of successfully adjusting to modernization. The Selva Lacandona region is one area where campesinos have experienced tremendous difficulty in adapting to the new global agriculture. Not only have agricultural transformations encouraged migration into the Selva Lacandona, but they also provide the context for the marginalization of most of the region's population. This context is better understood by examining the responses to agricultural changes in Chiapas, or in some cases the lack of responses.

Agricultural Politics in Chiapas

Mexico's economic liberalization policies have had a notable impact on Chiapas. For example, there has been a dramatic reduction in public investments in rural development: The amount invested by the federal government in 1991 amounted to a mere 8.7 percent of the 1982 value. The availability of credit through the Banrural, Mexico's rural development bank, has likewise contracted: The availability of credit for the production of the five most important crops shrunk from 1,700 million pesos (old) in 1987 to 285 million pesos (old) in 1992. Finally, the value of the state's three most important crops (maize, coffee, and cattle) decreased by 23, 84 and 52 percent, respectively, between 1987 and 1992 (García Aguilar and Villafuerte Solís 1996).

The changes in Mexican agriculture and their impacts on Chiapas must be considered within the context of agricultural politics and agrarian policies in Chiapas. Overlaying the two, one begins to understand the enormous pressures on the Selva Lacandona, including pressures both internal and external to the region. Although agricultural changes at the state level have been closely tied to national policies, the combination of labor exploitation, land appropriation, and political alliances that emerged within Chiapas distinguish it as a unique case within Mexico.

In the following sections, the evolution and tenacity of land concentration in Chiapas will be traced. The governing class, which has historically been synonymous with the landholding class, has continuously sought to preserve and perpetuate their interests at the expense of a growing population of landless and unemployed peasants. When land reforms became inevitable, redistributions were avoided by simply directing land grants toward the tropical forests of the Selva Lacandona. This satisfied both landowners as well as land-seeking peasants, but contributed to the loss of what is considered by many to be a unique and invaluable ecosystem.

The economy of Chiapas is based largely on agriculture. Chiapas ranks second among Mexican states in terms of overall agricultural productivity, and it is Mexico's largest producer of coffee. Nationally, Chiapas ranks third in the production of maize and cattle, and fifth in the production of beans (Table 5.1).[2] In contrast, Chiapas is one of the least industrialized states in Mexico. Most of the existing industries are, in fact, tied to agriculture. Although the economy has diversified in recent years as a result of investments in the energy and tourism sectors, agriculture remains the cornerstone of the economy, involving 60 percent of the economically active population.

TABLE 5.1 Agriculture of Chiapas in the national context.

Top 5 Producers by State			
State	Maize	State	Coffee
1 Jalisco	1,063,956	**1 Chiapas**	**645,650**
2 Guanajuato	869,831	2 Veracruz	624,758
3 Chiapas	**767,645**	3 Oaxaca	248,565
4 Veracruz	587,990	4 Puebla	211,355
5 Chihuahua	580,767	5 Nayarit	70,407
Total - Mexico	8,309,514	Total - Mexico	1,947,046
State	Beans	State	Cattle
1 Zacatecas	255,977	1 Veracruz	178,396
2 Durango	102,840	2 Jalisco	158,551
3 Chihuahua	83,608	**3 Chiapas**	**79,730**
4 Guanajuato	66,640	4 Chihuahua	69,247
5 Chiapas	**59,135**	5 Sonora	62,411
Total - Mexico	913,119	Total - Mexico	1,275,940

Note:Maize, beans, and coffee data are for 1991; cattle for 1993.
Sources: INEGI 1994; INEGI 1995.

The transformation of Mexican agricultural production toward a capitalist, export-oriented system led to the present-day agrarian structure in Chiapas, which is made up of relicts of Porfirian haciendas, agroexporting farms, large cattle ranches, small private landholdings, ejidos, and communal indigenous lands (Muench 1982). Throughout recent history, land use in Chiapas has been characterized by an extremely high concentration of land in the hands of a small number of owners. A large number of ejidos exist alongside these concentrated landholdings, but often on marginal lands. To understand how this came about, one must go back as far as the colonial period in Chiapas.

The Colonial Legacy

Since the colonial period, there has been a continuing tension within Chiapas concerning the orientation of agricultural production. The tensions emanated from the existence of a large indigenous population practicing traditional subsistence agriculture on communal lands. When the Spaniards first entered Chiapas in 1524, the state was populated by approximately 14 ethnic Indian groups divided among five provinces (Muench 1982). These groups included Chañabales, Chiapanecos, Choles, Lacandones, Mames, Mayas, Mexicanos, Quelenes, Quichees, Tojolabales, Tzeltales, Tzotziles, Xoquinoes, and Zoques. They practiced slash-and-burn agricultural techniques developed over centuries and adapted to the natural conditions of each particular area. Their principal crops included maize, beans, and squash, along with a number of local specialties.

The agricultural techniques of the Indians were looked down upon by colonial elites, who were concerned with extracting wealth from the land in order to send abroad, as well as with accumulating personal fortunes. The Spanish conquerors wasted little time in disrupting the social and economic organization of the indigenous communities, primarily through the creation of formal institutions such as the *encomienda* system and *repartimiento*.[3]

Indians were expected to contribute to the aggregation of colonial wealth, not only through tribute from portions of their harvests (the encomienda system), but also with manual labor extracted through a system of forced labor known as repartimiento. Since the advent of the colonial system in Chiapas, human labor has been considered a natural resource, and one which has been fully exploited to maximize agricultural production and trade of goods. Consequently, colonial haciendas and ranches engaged a growing part of the indigenous labor force, which through the repartimiento system was free or extremely cheap (Wasserstrom 1983). According to Marion Singer (1988:20), by the mid-17th century the majority of Zoque, Tzotzil and Tzeltal Indians were accustomed to walking great distances carrying merchandise for export markets in order to earn money for survival, since all of the maize, bean, and chile that they produced was given as tribute to the colonial administration.

Likewise, indigenous lands were incorporated into the colonial system. Although the Spanish colonists did not do away with communal lands entirely, many of the best lands were expropriated by haciendas owned by colonial administrators or the Church. Most of the indigenous groups located in peripheral areas were able to hold on to their land and productive forces until the onset of unification with Mexico, when they were dispossessed of them by a recast landowning class made up from remnants of the traditional elite.

The Spaniards disrupted not only the social and production systems of the indigenous groups, but also the geographic distribution of the populations. This was accomplished through resettlement programs. Indigenous groups from the tropical lowlands, including the Selva Lacandona, were moved to the Highlands, where the colonial administration and the large haciendas were located. In addition, much of the Indian population was decimated by foreign diseases (Wasserstrom 1983). The Highland population eventually recovered and increased in number, whereas the lowlands remained sparsely populated. The population decline among indigenous groups in Chiapas not only increased the demand for their labor, but also ensured a social stability which allowed the colonial system of labor exploitation to perpetuate itself (Marion Singer 1988).

Once Chiapas was unified with Mexico in 1824, the newly emerging group of *criollos* that replaced the colonial rulers enacted a series of agrarian laws that facilitated the transfer of the choicest Indian lands into the hands of non-Indian elites (Wasserstrom 1983).[4] These laws, passed between 1826 and 1832, allowed landowners to obtain title to village lands that were not under cultivation, and as a result the number of farms in Chiapas multiplied (Marion Singer 1988). The new landless Indians were not pushed into the tropical forest frontiers of Chiapas, as would be the case in the next century, but were instead appropriated by landowners as tenant or wage laborers to satisfy the growing demand for labor in Chiapas.

Agricultural production on haciendas and later on latifundios was based on a complex system of labor relations. In the post-colonial period, slavery, encomienda, serfdom, and tribute systems common to the colonial hacienda were replaced by or intermingled with share-cropping, tenant farming, and above all, debt peonage (Bartra 1993). The existence of these forms of labor attests to the fact that, in spite of the facility to acquire land, labor shortages continued to be one of the most serious problems for the rapidly growing class of landholders in Chiapas (Wasserstrom 1983; Benjamin 1989). According to Benjamin (1989:14), "[t]he struggle for the control of Indian labor, often disguised by the larger conflict over the power and privileges of the Church, was one of the central motivations of political strife in the decades following annexation [to Mexico]." Attempts to address the problem included the Servitude Law of 1827, which enlisted the labor of vagrants into the military or to fulfill the needs of labor-short employers. This did not, however, adequately resolve the problem, as "most of the potential laborers were Indians from the Highlands, whose labor was controlled by landowners and priests" (Benjamin 1989). Most of the large plantations, in contrast, were located in the lowlands.

The landowning class that came to dominate in Chiapas after independence was not homogeneous. Instead, it consisted of two distinct factions, known collectively as the *familia chiapaneca*. The first faction, concentrated in the Highlands and Sierra Madre of Chiapas, could be described as conservative and traditional. It was engaged in semi-feudal production systems oriented toward the local market and facilitated through the exploitative use of Indian labor. In contrast, a more liberal and progressive group of landowners emerged in the lowlands and Central Valley of Chiapas. This latter group was concerned with the expansion and modernization of agriculture, and favored the integration of Chiapas into the larger Mexican economy through the development of capitalist agriculture. The central political conflicts in Chiapas during the nineteenth and early twentieth century revolved around the tensions between the two factions, with indigenous groups often playing a peripheral role or serving as pawns in the regional rivalries. The emergence and development of this rivalry between the landowning *familia chiapaneca* is described by Benjamin (1989:1):

> Politics was not a matter of class or race but of economic geography. The surviving colonial oligarchy of the province — the clergy, landowners, and merchants located in the provincial capital in the Central Highlands — subsisted on the labor and the surplus production of the large nearby Indian populations whom the elites regarded as a "natural resource" . . . The farmers and merchants of the less populated but more fertile Central Valley coveted Indian labor and church lands and, therefore, provincial governmental power to implement "reforms" to transfer those resources into their more productive hands.

The tension between rival groups of landowners struggling to promote or protect their own interests was manifested in political maneuverings that involved shifting alliances with national governments. It was not until the Porfiriato period that the liberals of the Central Valley were able to successfully initiate the process of modernization and integration into the larger economy of Mexico. Beginning in 1877, a succession of governors of Chiapas were unconditional Porfiristas. According to Benjamin (1989:21), an "informal but effective system of patronage brought the periphery increasingly under the influence of the center." As a result, the once-weak state government grew considerably stronger. As a consequence of this new strength, elite landowners in the Central Valley sought to end the power maintained by local and regional *caciques* and their institutions. They sought a modernized

Chiapas, complete with infrastructure that would, in effect, enhance their own economies.

During the Porfiriato, the Central Valley became the most economically dynamic region of Chiapas. The main export products of Chiapas found new markets and better prices, and Chiapas experienced a period of great economic progress. Favorable external conditions for agricultural crops such as cotton, sugar, rice, and coffee made Chiapas an important supplier to Mexican and North American markets (Benjamin 1989). At the same time, foreign investments flowed increasingly into Chiapas, as European and North American entrepreneurs established plantations in crops and materials such as coffee and rubber. Nowhere was this more important than in the Soconusco region of southern Chiapas, where coffee production had been introduced at the end of the nineteenth century from Guatemala. German immigrants set up an extensive system of coffee production on land that they had acquired from surveying companies for pennies per hectare. This enclave economy alone provided the impetus for economic growth in Chiapas during the Porfiriato (Benjamin 1989).

As described in the previous chapter, the Selva Lacandona during this period was a source of prosperity for owners of logging firms and international capitalists involved in the tropical timber trade. It was also a dynamic center for rubber production, particularly around Palenque (Wasserstrom 1983). However, the thick vegetation and mountainous terrain made the core of the Selva difficult to penetrate, thus unattractive for capitalist agriculture. It would not be until decades later that access roads or sheer urgency would enable the penetration of agriculture into the remote area of the Selva Lacandona.

Land Concentration in Chiapas

Land ownership in Chiapas was firmly consolidated in the hands of a select few by the first decade of the twentieth century. The same laws which had allowed timber and surveying companies to acquire vast landholdings in the Selva Lacandona enabled individuals to accumulate expansive landholdings throughout Chiapas, in the name of capitalist development and free enterprise. Between 1889 and 1908, the number of latifundios in Chiapas tripled from 3,159 to 8,527 (Reyes Ramos 1992). The Chiapas economy experienced a tremendous rate of growth during the same period. However, the benefits of modernization witnessed during this period did not trickle down to the majority of the population in Chiapas. Instead, modernization was largely to the advantage of landowners and merchants.

While national laws enabled individuals to amass large estates, the Chiapas government initiated its own reforms in an effort to modernize the state economy. Governor Emilio Rabasa (1891-1894) implemented a land reform that, while considered an economic success, further alienated Indians from their traditional lands (Benjamin 1989:48-49):

> Rabasa strongly believed that the division of communal village lands and the creation of a new class of yeoman farmers would promote productive capitalist agriculture and the integration of the Indian (and traditional campesino) into Mexican civilization. To advance these goals Rabasa enacted and vigorously enforced a measure . . . to divide ejidos in Chiapas into private parcels. . . . Enterprising sharecroppers, renters, small merchants, and ranch foremen benefitted most from this opportunity to become landowners. . . . The effect of the reparto upon Chiapas's villages, however, was devastating. . . . As the number of ranchos and haciendas increased, communities that had been independent for hundreds of years either disappeared or became hacienda rancherias.

One immediate consequence of Rabasa's land reform was that more villagers were forced into migrant labor, indebted servitude, sharecropping, and tenant farming. By 1910, out of a total population of 400,000, somewhere between 75,000 and 150,000 were agricultural workers. Of these, one-third to one-half worked as indebted servants, while the remaining worked as day wage laborers, sharecroppers, or renters (Benjamin 1989). According to Wasserstrom (1983:151):

> By 1910, it was clear that Liberals in Chiapas had accomplished one of their most cherished goals: they had drastically altered both the region's economy and the social relations upon which that economy was based. To this end, they had effectively deprived most native communities of their land, transformed many highland families into lowland fieldhands, and compelled the Indians who remained in their towns to pay rent and the capitación [tax].

In the indigenous communities of the Central Highlands, entire villages were tied to farms, and a large number of Indians migrated to the coffee plantations during the harvest season.[5] However, many Indians still lived on communal lands in the Highlands, as the Rabasa reforms had affected this area the least. In fact, some indigenous groups had managed to retain control of their land throughout extensive periods of land alienation and expropriation. These were communities which were outside the reach of the expanding belt of capitalist farming. Nevertheless, they could not compete with the larger farms and plantations in the regional market, thus their products were devalued and they were forced to occasionally hire out their labor to larger haciendas and farms (Marion Singer 1988). Conditions on these communal lands were ever deteriorating, as populations grew and land availability stagnated or decreased.

Although the same conditions would later drive peasants into the Selva Lacandona, at the time this was not an option. Most workers were not, in fact, free to migrate. In the Highlands, landowners not only maintained access to labor through debt, but also through the development of paternalistic relationships which tied laborers to a particular farm and landowner through what resembled feudal patron-client associations. Landowners served as owners, advisors, and tutors, controlling all aspects of the workers' slave-like existence (Marion Singer 1988). These ties softened the appearance of exploitation, but hardly changed the reality of it (Grindle 1986). The ties were often so dominant that during times of political conflict, workers often fought *for* rather than *against* the landowner.

These paternalistic ties became extremely significant during the Mexican Revolution. In Chiapas, the revolution acquired character-istics completely different from other parts of the country. The revolution did not arrive as a social movement, but instead was seen as the imposition of obligations by those who had gained political power in the federal government in 1914. The armed struggle in Chiapas was therefore not between peasants and landowners, but between land-owners and federal forces. In fact, the rivalry between the two factions of landowners subsided during the period of the Mexican Revolution (Benjamin 1989). Faced with an external threat to their power through nationalist policies aimed at breaking up large landholdings and ending archaic forms of labor exploitation, many of the Liberals and Conservatives joined together to create a counterrevolutionary force. The landowners united against the Carrancista government to regain state power in order to maintain the privileges they had acquired over the past centuries (García de León 1984b; Benjamin 1989).

To the campesinos, the social movement of the revolution was perceived as external and distant. As a result, the majority did not participate in the struggle, and the few who did were allied with the landowners, in defense of what had until then been their basic means of sustenance — the hacienda (Reyes Ramos 1992). In contrast to this, the Carrancista military governor of Chiapas, General Jesus Augustín Castro, sought to enact revolutionary changes to the appalling social conditions in Chiapas. Castro proclaimed a number of laws and decrees which "sought to liberate workers, small property owners, Indians, women, and municipal governments from the control of economic, spiritual, political, and domestic bosses" (Benjamin 1989:121). Nonetheless, the final result was that "Chiapas experienced the 'effects of the Revolution,' but was not itself revolutionized" (Benjamin 1989:119). The fact that the Mexican Revolution essentially bypassed Chiapas had great implications for the future of agrarian policy, as well as for social organization within the state. The failure of the revolution to break up large estates meant that in order to accommodate future demands for land, the agricultural frontier would have to be expanded into new areas.

The assassination of President Carranza in 1920 led to the triumph of the counterrevolutionaries in Chiapas (see García de León 1984b and Benjamin 1989). The landowning counterrevolutionary force did not have a specific political program, other than to reinforce their own privileges and protect their estates against national reforms. As a result, anti-agrarian policies were maintained. The legislation that was passed in Chiapas during the key years for Mexican agrarian reform (1914-1940) led to the present-day land tenure and agrarian structure (Reyes Ramos 1992).

Post-Revolutionary Agrarian Reforms

Numerous agrarian laws were passed in Mexico during the post-revolutionary period. However, the content and effect of these laws depended largely on contemporary social forces and on the conception of agrarian reform maintained by each successive government (Reyes Ramos 1992). As a result, many of the laws seemed contradictory. The constant shifts and alterations in federal government policies allowed for a wide margin of actions to be taken at the state level, which made it possible for the government of Chiapas to pursue its own type of agrarian policy (Reyes Ramos 1992). Consequently, the landowning class in power only superficially embraced the land reform policies adopted by national governments.

In 1921, Chiapas governor Tiburcio Fernández Ruiz, a landowner, issued the *Ley Agraria del Estado*, with the fundamental objective of preserving latifundios. This law established the maximum size for private property as 8,000 hectares, which was one of the highest ceilings in Mexico. It also offered landowners the possibility of dividing their lands and selling off parcels. In this manner, the state avoided breaking up the large holdings and giving them away to campesinos. Instead of creating transfers between landowners and peasants, the law simply established land as a commodity for purchase or sale, protecting the landowning class (Reyes Ramos 1992). Meanwhile, in 1922 the national government ruled that land used for plantation agriculture, including coffee, cacao, vanilla, and rubber, were not to be included in agrarian reforms. This was extremely significant for Chiapas, as the largest of landholdings were dedicated to coffee production. With such measures, the state policies protecting landowners were reinforced by the federal government (Reyes Ramos 1992).

Agrarian reforms in Chiapas during the decade following the revolution were extremely weak (Reyes Ramos 1992). Between 1920 and 1929, 31 agrarian acts were passed which gave 46,607 hectares to 5,026 campesinos, primarily in the Soconusco region. In part, these reforms were enacted to overcome the labor shortage caused by the 1914 *Ley de Obreros*, which prohibited servitude in Chiapas (Reyes Ramos 1992). The earliest reforms gave out land bordering on the coffee plantations, with the purpose of attaching the campesinos to the plantation as a supply of labor, property-owning nonetheless (Reyes Ramos 1992). A policy of selective reform emerged, whereby state lands or properties abandoned by foreigners were distributed with the intention of transforming landless peons into less militant ejidatarios (Wasserstrom 1983). Yet by 1930, the land tenure situation in Chiapas remained highly concentrated:

> There were 29 fincas that possessed more than 10,000 hectares; together they held more land (roughly 900,000 hectares) than the 15,000 properties of 500 hectares or less (roughly 760,000 hectares). The 1,500 fincas of 500 or more hectares possessed 79 percent of all land; the 15,000 properties smaller than 500 hectares possessed 18 percent; and the 67 ejidos listed in the 1930 census possessed only 3 percent. Agrarian reform, it seems, had barely touched Chiapas (Benjamin 1989:179).

To understand why so little land was distributed to campesinos prior to 1934, two points must be considered. First, at the start of the Mexican Revolution, 92.8 percent of the agricultural population in Chiapas consisted of *peones acasillados*, or wage laborers. The earliest agrarian legislation did not consider wage laborers as eligible to submit land claims in their home areas. In other words, they had no right to solicit land from the haciendas where their families had worked for generations. Even though obstacles to land claims by peones acasillados were removed by national legislation in 1934, the system itself did not change. The state government lacked the political will to alter the haciendas, and instead responded by distributing land in remote zones, away from the petitioner's original home (Reyes Ramos 1992).

Second, the post-revolutionary demand for land in Chiapas was weak in the first place. Unlike most other states in Mexico, the campesino population of Chiapas was held together tightly by the social dominion of the estate, and bound by a complex social structure to the landowning class, such as clientelism and paternalism. As such, campesinos conceived of social reforms in terms of improving labor conditions on the landowner's estate, rather than in terms of land distributions (Reyes Ramos 1992).

The first phase of meaningful agrarian reform in Chiapas occurred between 1930 and 1940. This was a direct result of national politics. During the Cárdenas administration, the *Código Agrario* of 1934 made agrarian legislation the exclusive domain of the federal government. Consequently, national legislators were able to restrict the ability of state governments to define their own laws, and hence oblige the states to embrace agrarian reforms (Reyes Ramos 1992). Forced to comply with national policy, Chiapas undertook its first important agrarian reforms. These reforms coincided with a period when union activities were threatening to limit the power of the landholding class. The early reforms therefore served in part to weaken the communist and socialist movements that were starting to gain power in the state (Reyes Ramos 1992).

By the 1930s, the communist labor movement in Chiapas was, in fact, a serious political threat to the establishment. The mobilization of workers and campesinos into socialist and agrarian organizations emerged as a movement during the 1920s. Communist groups, unions, agrarian committees, and socialist defense leagues had formed, with their actions directed largely toward the most developed regions of the state (Reyes Ramos 1992). The newly established *Partido Socialista Chiapaneco* helped bring socialist governor Carlos Vidal into power in 1925. Under the appearance of radical socialism, Vidal followed "a blueprint for state capitalism via a populist brand of bourgeois revolution" (Benjamin 1989:163). Nevertheless, Vidal also initiated the first serious and populist land redistribution program in Chiapas.

The Chiapas government responded with attempts to unify and control the labor movement within the state. It did so through a number of means, including the creation of the *Confederación Campesino y Obrera de Chiapas* (CCOC) in 1931. This organization, and others that followed, offered the government a means of controlling organized rural farm labor, establishing a corporatist tactic that would serve landowners well in the future:

Labor federations and agrarian leagues were incorporated into the political apparatus of the state, their leaders were transformed into

interchangeable politicians, and their constituent unions and ejidos sank
into fratricidal struggles for short term gain. Reformers were transformed,
by means of the corrupting influence of 'politics,' into instruments of
pacification, manipulation and control (Benjamin 1989:195-196).

Landowners did not just sit quietly and let the national policies of
Cárdenas slowly erode their privileges. Instead, they organized in
opposition, largely though the formation of Cattlemen's Associations.
A 1940 state law permitted the formation of private police forces to
patrol cattle lands, thus creating the infamous "white guards" which
have since then repressed any efforts among peasants to organize or
take over land (Reyes Ramos 1992). Regardless of sympathetic state
and federal government policies, the desire to create ejidal lands
initiated a violent struggle between landowners and peasants. Landowners attacked agrarian communities, burned settlements, and
murdered peasant activists (Benjamin 1989).

The Backdrop to Colonization

Fundamental changes have occurred in Mexican agriculture over the
past five decades. The role of peasant production systems within
Mexico's agricultural sector has become increasingly irrelevant, and the
outcome has been a profound crisis in rural Mexico. These changes serve
as the backdrop to colonization of the Selva Lacandona, which will be
discussed in the following chapter.

The backdrop to colonization is also related to the tenacity of
outdated agrarian relations in Chiapas. The precedents to the land
reforms of the post-1940 period in Chiapas were described in detail in
order to lay out a number of points. First, land ownership was concentrated in the hands of relatively few prior to 1910, and this
distribution was not altered by the Mexican Revolution. Second, the
governing class of Chiapas has historically been equivalent to the
landowning class, which has governed almost exclusively to maintain
its own interests. Third, national policies enacted in the 1930s, along
with the rise of socialist parties, labor unions, and agrarian
committees, made it increasingly difficult for the government of
Chiapas to avoid land reforms. Fourth, by institutionalizing labor and
agrarian movements, the government of Chiapas was able to maintain
control over the countryside and direct land reforms in the manner most
suitable to it. Finally, by acquiring the right to maintain private police
forces, landowners were ensured a means of keeping peasants in
compliance with their wishes. These factors set the stage for an

extensive land distribution and colonization program that began in the 1940s. The lands sacrificed to this program were not found on the big estates, but rather on national lands, including the Selva Lacandona.

6

Colonizing the Selva

Why have colonizers and private landowners moved into the remote, humid, tropical forest environment of the Selva Lacandona? Moreover, why has cattle ranching emerged as a desirable mode of production among both private ranchers and small-scale farmers? Over the past twenty years, numerous studies have been undertaken in an attempt to understand both the colonization process and the agricultural systems that have emerged in the once-forested areas of the Selva Lacandona (Lobato 1980; Price and Hall 1983; Mauricio Leguizamo et al. 1985; Garza Caligaris and Paz Salinas 1986; Paz Salinas 1989; Leyva Solano and Ascencio Franco 1996). Most of these studies have focused on one or several communities in a particular region, and many include a comprehensive analysis of the social and economic structures which have evolved. These works contribute to a broader understanding of the many changes that have been taking place in the Selva Lacandona. They also provide insights to the patterns of deforestation that have resulted from these changes.

Land Reforms in Chiapas

Agrarian reforms had little impact on Chiapas in the years following the Mexican Revolution. By 1940, land was still highly concentrated. More than half the land remained in the hands of only 2.6 percent of all landowners in Chiapas, despite the fact that land distributions had been carried out to the benefit of 25,644 campesinos. This group, representing the vast majority of all landowners (76.97 percent), possessed a mere 4.39 percent of the land (Reyes Ramos 1992). One repercussion of this skewed distribution was stagnation in the agricultural sector of Chiapas.

Between 1930 and 1940, the total area harvested had only grown from 167,501 hectares to 184,932 hectares (Reyes Ramos 1992).

This stagnation could be attributed to the lack of diversified products, and to a prevailing tendency for production to be aimed at on-farm consumption rather than the market. Out of 12,581 properties greater than 5 hectares, 7,240 produced maize and beans for subsistence (Reyes Ramos 1992). The engines of growth in agricultural production were limited to the highly concentrated export enclaves, which focused production on a few specific commercial crops, such as coffee, cacao, and bananas.

Land itself was not a scarce resource in Chiapas, but it was underutilized. Low agricultural productivity could partly be attributed to absenteeism among landowners. Some 2,569 properties (accounting for 1,314,125 hectares) were in the hands of administrators rather than landowners (Reyes Ramos 1992). For many estate owners, the hacienda was considered nothing more than a source of perpetual rent, and at times, a place for a seignorial excursion (Bartra 1993).

The state government pursued several strategies to address agricultural stagnation. For example, it first tried to encourage agricultural production on unused or underutilized lands. This was achieved through a variety of legislation, including the 1939 *Reglamento de la Ley de Tierras Ociosas*, or the law of underutilized lands. Such measures did not, however, remedy the situation, nor did they alleviate the growing demands for land among the landless of Chiapas.

Finally, the state government embarked on an extensive agrarian reform program. Between 1940 and 1949, land was distributed in almost three-quarters of Chiapas (74 municipios). Yet almost half of this land was concentrated in only 11 municipios. These distributions were primarily located in the Soconusco region and along the border with Guatemala. This was not accidental, for it ensured a supply of labor to the coffee plantations and served to define the border with Guatemala (Reyes Ramos 1992).

Thiesenhusen (1989) stresses that true agrarian reform involves more than simply land reform. He points out that "other institutions must be redirected and reshaped at the same time that land is redistributed to insure that services, inputs, research, irrigation water and facilities, credit, and marketing assistance go to the beneficiaries of land reform" (Thiesenhusen 1989:7). According to Reyes Ramos (1992), the distributions undertaken in Chiapas were not representative of true agrarian reform in that they did not involve a redistribution of both land and income, including modifications in the economy as well as transformations in social and political structures of the region or country. Land distributions in Chiapas rarely involved the breakup of large estates, and existing social and political structures were untouched. In fact, the government simultaneously worked to ensure that private property rights in the countryside remained intact. Local economies tied to cattle and export agriculture were protected from reforms through a number of legislative measures, including exemption certificates (*certificados de inafectabilidad*).

A somewhat contradictory scenario emerged in Chiapas, where extensive amounts of land were distributed to campesinos while large estates remained intact. This contradiction can be explained by the fact that most of the land distributed over the past 50 years has been national territory or previously unsettled areas. In 1940, Chiapas possessed about 3 million hectares of such national lands (Reyes Ramos 1992). New areas, particularly in the Selva Lacandona, were opened up for colonization in order to divert attention away from the source of agrarian tensions, namely land concentration (García de León 1984b). The population of the Selva Lacandona region increased dramatically over a period of thirty years, approaching 300,000 in 1990 (Figure 6.1).

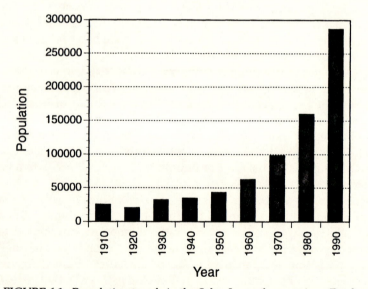

FIGURE 6.1 Population trends in the Selva Lacandona region. Totals are based on the populations of Ocosingo, Las Margaritas, Altamirano and Palenque (*Source*: Ascencio Franco & Leyva Solano, 1992).

Colonizing the Frontier

Colonization of national lands had been a long-established policy of the Mexican government, and it tended to favor private property over communal forms of land tenure.[1] This led to private, large-scale land accumulation on Mexico's national lands. In Chiapas, 171 individuals, or *nacionaleros*, occupied and solicited title to 278,084 hectares of land prior to 1934, averaging 1,626 hectares per individual. Ejidatarios who were titled with national lands, on the other hand, received an average of only about 9.27 hectares each (Reyes Ramos 1992). The preference for granting state lands to individuals remained intact until 1946. At that time, a new *Ley Federal de Colonización* was signed by Mexican president Miguel Alemán, giving priority to campesinos who were most in need of land. It fixed the size of new land grants neither to exceed that of small properties, nor be smaller than that of an ejidal plot.

Despite the new federal colonization law, individuals continued to solicit and receive private parcels throughout Mexico, and a great deal of land speculation resulted. In fact, the best lands were often given to nacionaleros (Reyes Ramos 1992). By the time the law was remedied in 1962, almost a half million hectares of land were awaiting title to nacionaleros in Chiapas. The 1962 law protected these and other already established private properties from expropriation. However, it also stipulated that future colonization would only be authorized for the creation of ejidos, through what became known as *Nuevos Centros de Población Ejidal* (NCPE) (Reyes Ramos 1992; Paz Salinas 1989).

The colonization of the Selva Lacandona through NCPEs represented an attempt to control spontaneous settlements in the region. While ejidal lands in the Selva Lacandona were typically selected and colonized by peasants *prior* to petitioning the government for title to the land, NCPEs were assigned by government decree and then settled by colonizers (Neubauer 1997). The size of NCPEs was also larger than traditional ejidos. Each ejidatario was entitled to 50 hectares, of which 20 were designated for individual use and 30 were to be managed collectively by the ejido.[2] Between 1960 and 1984, 219,334 hectares of land in Chiapas were distributed through 83 NCPEs, benefitting 6,154 ejidatarios (Reyes Ramos 1992). Of these, 14 were located in Ocosingo, and only two in Las Margaritas.

The large quantity of national lands in Chiapas gave the state government a wide margin within which to manage land distribution policies. It meant that it was not necessary to disturb the interests of landholders to satisfy the demand for land by the campesinos.

Consequently, landowners maintained power and did not disappear as a hegemonic class, as they did in other parts of the country (Reyes Ramos 1992). Furthermore, the creation of NCPEs provided the Mexican government with a means of resolving land requests in areas where no land was available for distribution, or where the granting of land would involve political conflicts with landowners. With NCPEs, campesinos could be granted land not only in other municipios, but also in other states (Reyes Ramos 1992).

Although colonization was initiated in the 1940s, it was not until the 1950s and 1960s when the municipios of the Selva Lacandona region were opened up for massive colonization. Some of the first groups to colonize the Selva Lacandona during the 1950s were Tzeltal Indians from Bachajón who were fleeing social conflicts created by land tenure disputes (Dichtl 1987). Tzotzil Indians from Candelaria and Baxek'en in the Highlands were among the first to settle along the Río Santo Domingo in Las Margaritas, creating a port of entry for numerous other groups that followed (Garza Caligaris and Paz Salinas 1986). During the 1950s, 82,638 hectares were distributed in the municipios of Ocosingo, Las Margaritas, and Altamirano (Reyes Ramos 1992). In the following decade, an additional 83,449 hectares were given out in the municipios of Ocosingo and Las Margaritas alone (Reyes Ramos 1992). These distributions took the form of ejidal land grands, NCPEs, community lands that were granted or restored to indigenous groups (*bienes comunales*), and amplifications of ejidal land grants.

Early colonization of the Selva Lacandona was entirely spontaneous. The government did not indicate where settlements should be located, nor did they provide any type of support to the migrants. When the government did get involved in the colonization process, it was largely to handle the paperwork associated with land tenure claims. Initially, colonization spread slowly, with the best lands settled first. Most of the fertile lands in the valleys of Las Margaritas and Ocosingo had already been claimed by nacionaleros for private farms, or *fincas*. Often, workers would abandon the fincas in search of their own land, and they usually ended up settling nearby.[3] Meanwhile, groups of immigrants, usually consisting of family and friends, were also heading to the forest in search of land. Lobato (1980:29, own translation) described the process in the following manner:

> After seeing the possibilities for land distributions, the first isolated families that had settled in the forest returned to their original homes to look for new friends, acquaintances, relatives, or simply anyone who had the desire to go to the forest; in this manner, new immigrants continued to

arrive from far-away places; some of them were able to join the ejidos as members, others remained nothing more than residents, that is, campesinos without ejidal plots.

As colonization continued, only the most marginal lands were left for newcomers, unless, of course, one was willing to venture into more remote areas. The colonization process inherently led to social stratification within the communities of the Selva Lacandona (see Preciado Llamas 1976 and Lobato 1980). The earliest settlers were often able to appropriate the labor of recent settlers who were in the process of acquiring land or establishing subsistence. As such, they became the first to acquire capital and venture into commercial activities. Rather than accepting marginal land, many of the migrants soon relocated further into the forest, where they could take the role of founding settlers and lay claim to the best land.

By 1961, the implications of spontaneous colonization were becoming clear to the government. More and more people were moving into the Selva Lacandona, and the government had no control over the process. Consequently, several institutions were charged with organizing settlements. Two institutions in particular played significant roles in promoting colonization: the *Departamento de Asuntos Agrarios y Colonización* (DAAC) and the *Instituto Nacional Indígena* (INI). In an effort to satisfy the growing demand for land among campesinos in the northern part of Mexico, in 1961 DAAC proposed the formation of 17 settlements in the Selva Lacandona based on private land tenure (Paz Salinas 1989). However, the land destined for these settlements was already under colonization by indigenous groups from the Highlands. In the end, only four out of the projected seventeen settlements were actually formed. The fact that spontaneous colonization was already underway rendered the government's belated intervention more or less ineffective.

Another attempt to manage settlement in the Selva Lacandona was initiated in 1965. INI, in collaboration with DAAC, established a directed settlement program aimed at populating 60,000 hectares of national lands in the municipios of Las Margaritas and Independencia, in the southeastern part of the forest. The program involved an attempt to resettle 10,000 families, or about 50,000 people, from the densely populated Highlands of Chiapas into the tropical forest. This reflected INI's perception of the situation in the Highlands of Chiapas: poverty and land tenure problems were seen as entirely demographic issues, which could be solved by expanding the agricultural frontier onto national lands and relocating the excess population (Garza Caligaris and Paz Salinas 1986; Paz Salinas 1989).

Population growth is frequently held to be the driving force behind colonization of the Selva Lacandona. Yet the agrarian history of Chiapas makes it clear that the situation is *not* simply an outcome of indigenous people from the Highlands reproducing faster than the land can accommodate them. Instead, colonization reflects historical structures brought about by agricultural transformations and the ability of a landowning elite to amass and preserve large estates, at the same time maintaining access to a cheap labor force. A study by Wasserstrom (1983:201) which examined class formations and social relations within two municipios in the Central Highlands of Chiapas demonstrates this extremely well:

> As we know, for nearly a century state authorities and private landowners alike have collaborated to transform the [Highlands] into a source of abundant and ill-paid seasonal laborers. In order to assure that such conditions prevailed after land reform, they permitted municipal authorities (particularly former scribes) to acquire large tracts of communal land and to amass extensive private estates. In turn, these men, who dominated the town's agrarian committees, allowed a few favored allies in each hamlet to accumulate considerable resources of their own.... By lending money at usurious rates, they soon extended their hold over most of the community's productive farmlands and all of its commerce. As a result, even those men and women who became ejidatarios in 1937 and 1938 were quickly forced to rely upon wage labor as their primary source of income. By 1970, in fact, such conditions had become the common lot of over three-quarters of the municipio's population.

The resettlement solution proposed by INI did not address these issues. However, it did include numerous economic development programs, such as commercial agriculture, cattle ranching, and forestry. More important, a road was planned that would connect the ejidos to the regional market at Comitán.[4] INI's program was scheduled to begin in 1966. However, like many government projects initiated in the Selva Lacandona, the plans were never realized. In reality, the government did little more than promote the colonization scheme, leaving settlers to cope for themselves (Garza Caligaris and Paz Salinas 1986; Paz Salinas 1989).

The Mexican government also indirectly promoted colonization of the Selva Lacandona through its social policy toward indigenous populations. During the 1950s, the idea of acculturation was widely believed to be the solution to the underdevelopment of indigenous groups. A controversial Protestant sect concerned with the acculturation of Indians had been invited into the country by president Lázaro Cárdenas. The group, known as the Wycliffe Bible Translators or the Summer Linguistics Institute, invested vast amounts of human and financial resources to promote migrations from the Highlands to the

Selva, where it would be easier for them to convert uprooted Indians to their evangelical religion (Dichtl 1987).

Another religious group, the Jesuits of the Catholic Church's Misión de Bachajón, played an important but very different role in the lives of indigenous populations beginning in the 1970s. Rather than acculturation, these missionaries sought to rescue and preserve valuable cultural traditions as well as organize work cooperatives for the benefit of the communities. Religious conflicts in Highland communities also pushed groups into the Selva Lacandona. In some cases, Protestants were expelled from their communities, in other cases, they left on their own in search of "the promised land." During the 1980s, the Seventh Day Adventists also established missionaries in the Selva Lacandona, converting a number of villages, including the Lacandones of Metzabok, to Christianity (Dichtl 1987).

Numerous oral histories tell of the difficulties that new settlers had coping in their new environment (see Garza Caligaris 1986; de Vos 1988b; Paz Salinas 1989; Townsend et al. 1996). They lived amidst a tropical forest full of strange wildlife, unknown fruits, abundant insects, and seemingly never-ending rains. Most were also cut off from their social networks and access to the marketplace. The Selva Lacandona represented an ominous fate for some, and it was not infrequent that colonizers turned around and went back to their original homes. However, many had nothing to return to, hence they persevered and "tamed" the forest. If their new homes were on national territory, then the land was cultivated for the required number of years before land tenure could be formalized. If the land had been previously settled by others, colonizers immediately applied for ejido status. Finally, if a newcomer had sufficient resources, land was purchased directly and a *ranchería* was formed. It was not uncommon for one or two families to detach themselves from the ejido and purchase land to cultivate maize and raise livestock, becoming satellites of the ejidos (Ascencio Franco and Leyva Solano 1992).

By the late 1960s and early 1970s, many of the ejidos in the Selva Lacandona had grown into villages. Between 1962 and 1975, at least 120 colonies were formed, and about 50,000 people had moved into the region (Price and Hall 1983). Many had been informed of the availability of land by friends or relatives who had already settled there. As a result, colonization was characterized by continuous waves of individuals and families that moved into different areas in search of land. However, eventually the best lands had been claimed, and existing ejidos were no longer accepting new members. New migrants were left to work as wage labor on established ejidos, while at the same time they searched for their own land to clear for cultivation (Paz Salinas 1989).

The pace of land distributions in the Selva Lacandona continued during the 1970s and 1980s. During the 1970s, 171,230 hectares of land were distributed in Ocosingo, and 21,406 hectares in Las Margaritas. In the 1980s, Ocosingo continued to be the leading municipio for land distributions. By 1984, at least 403,481 hectares had been distributed in the municipios of Ocosingo, Las Margaritas and Altamirano, which make up most of the Selva Lacandona region (Table 6.1). Including the 614,321 hectares granted to the Comunidad Lacandona, the amount of land distributed in the Selva Lacandona by 1984 exceeded one million hectares. Many of the most recent distributions were in remote parts of the forest, including the Marqués de Comillas subregion. This subregion was earmarked for colonization by the Mexican government as early as 1967, with the first settlers arriving in 1972 (de Vos 1988b). However, by 1980, it was still considered a distant frontier.

The availability of land in the Marqués de Comillas region, particularly along the Mexico-Guatemala border, coincided with a new wave of colonization that began in 1980 (González Ponciano 1990). Settlements in the Marqués de Comillas region expanded, facilitated by the roads constructed by PEMEX and reinforced by labor provided by Guatemalan refugees. Many of the newcomers to this area came from other impoverished Mexican states. The sources of migration included, in order of importance; Tabasco, Veracruz, Oaxaca, Campeche, Guerrero, Puebla, Distrito Federal, Michoacán, Yucatán, Estado de México, and Quintana Roo (Ascencio Franco and Leyva Solano 1992).

By the 1990s, the Selva Lacandona had been colonized on all fronts (Figure 6.2). Colonization generally followed valley floors, rivers, and roads. Nevertheless, as this land became increasingly scarce, new settlements emerged along ridges, and in other remote areas. The patterns of deforestation have led to a situation where the remaining forests lie in protected areas in the eastern part of the region. However, the pressures on these areas are intense. In fact, the interface between protected areas and settlements is where conflicts between environmental and social struggles are most visible.

The distribution of settlements as of 1990 is portrayed in Figure 6.3. The black squares represent population centers. The land surrounding these centers also belongs to the ejidos, communities, or ranches, such that little land remains available for distribution. In the Marqués de Comillas region, the size of ejidos is much larger than in the western and northern parts of the region, resulting in less concentrated communities. The map shows that what was once an "unknown desert

TABLE 6.1   Land distributions in the Selva Lacandona region, 1940 - 1984.

Municipio	Total Area (hectares)	1940-1949 (hectares)	1950-1959 (hectares)	1960-1969 (hectares)	1970-1979 (hectares)	1980-1984 (hectares)	Total Dist. (hectares)
Altamirano	112,000	–	16,990	–	–	–	16,990
Las Margaritas	571,800	19,959	41,838	43,643	21,406	53,514	180,360
Ocosingo	1,069,100	–	23,810	39,806	49,749	92,766	206,131
TOTAL	1,752,900	19,959	82,638	83,449	71,155	146,280	403,481
Chiapas	7,521,000	483,998	675,196	500,347	599,265	500,521	2,759,327

Note: A dash signifies that data were not available.
Sources : Gobierno del Estado de Chiapas, 1988; Reyes Ramos, 1992.

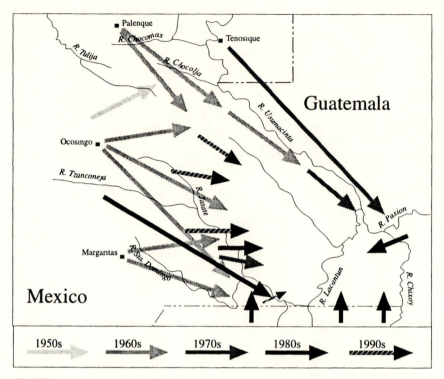

FIGURE 6.2 Patterns of colonization in the Selva Lacandona (*Source*: Modified from de Vos, 1988b).

inhabited by Lacandón Indians" is now an agricultural frontier that has absorbed thousands of landless peasants.

The situation of campesinos in the Selva Lacandona remains precarious, as exposed by the 1994 Zapatista Uprising. The marginalization of the population can be explained by considering the agricultural systems that have developed in this region, and their relation to trends in Mexican agriculture described in the previous chapter.

Farming Systems in the Selva Lacandona

A common stereotype applied to tropical forest ecosystems holds that their fragile soils cannot sustain agriculture. A more honest appraisal of the forest's potential for agriculture might acknowledge that the quality of land varies tremendously, and that some areas can prove to be quite productive if appropriate techniques are used, includ-

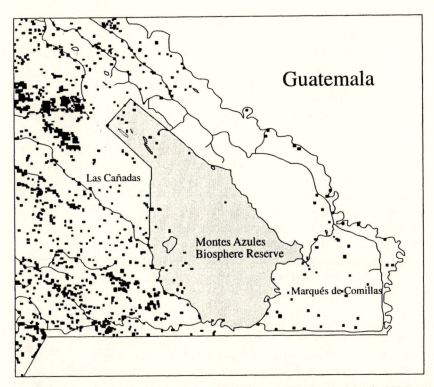

FIGURE 6.3 Settlements in the Selva Lacandona, 1990 (*Source*: March, 1994).

ing the proper use of fallows (see Ewel 1986). Much of the variation is attributable to the soils and terrain of the region. Land situated on mountain slopes and in seasonally inundated areas are of a lower quality than land located in well-drained valley bottoms.

Migrants from the Highlands of Chiapas and from other states are often accused of importing production systems unsuitable to a humid tropical environment. In contrast, many point to the traditional food production system of the Lacandón Indians as an ecologically sound alternative to the more destructive practices which have come to characterize the forest (Nations and Nigh 1980; Nations 1981). In reality, the production systems of the migrants have evolved to meet the demands of local conditions: "Production strategies tend to be designed to minimize the risk of crop failure and hunger through intercropping and enterprise diversity. Resource allocation decisions are made using the farmer's experience and knowledge of the performance of crops in the local environment" (Price and Hall 1983:58).

The farming systems that developed among the migrants to the Selva Lacandona can be considered dynamic and capitalist, despite reliance on rudimentary tools and human labor (Lobato 1980). Farming practices vary within communities and regions, and typically include a combination of subsistence agriculture, semi-commercial agriculture, and commercial agriculture. The dominant form of agricultural production is slash-and-burn cultivation. In the tropical environment, two crops can be grown each year: the major crop is planted before the summer rains begin, and the minor crop is planted during the dry season. The size of the area cleared for agriculture varies, but the regional average was between two and three hectares per family in the early 1980s. According to Price and Hall (1983:60), "[f]actors affecting milpa size include length of occupation of the farm, the quality of land, family size, labor availability, and enterprise mix." The production mix in the Selva Lacandona includes a combination of maize, beans, squash, bananas and plantains, coffee, chile, cacao, and sesame, as well as livestock production.

Although the primary objective is to maintain subsistence, most farmers strive to produce a surplus. In 1990, two-thirds of the maize produced by ejidos and smallholders was sold in the market, while the remaining third went to family consumption (Harvey 1996). Indeed, maize is the most important crop for subsistence purposes, and most other cropping systems revolve around its cycle. Clearing the forest to establish plots of maize is considered to be one of the most difficult tasks associated with farming. At the same time, it also helps to establish ownership of the land (Mauricio Leguizamo et al. 1985). Maize is grown in both intercrops and monocultures, and purchased inputs are rarely used. The largest part of the harvest is saved for on-farm consumption, but most farmers sell some of their harvest in average years (Price and Hall 1983). Although maize is usually the first commercial crop produced on a farm, it is not considered an attractive crop for commercialization because yields are uncertain, the price is not rewarding in relation to the effort involved, and it does not store well in the humid tropical climate (Price and Hall 1983).

During the early 1980s, about half of the maize sold commercially in the Selva Lacandona was marketed through a government-run agency, the *Compañía Nacional de Subsistencias Populares* (CONASUPO), while the remainder was sold to middlemen or marketed directly in nearby villages or towns (Price and Hall 1983). Although CONASUPO offered state-guaranteed prices, transporting the crops to government granaries and meeting quality standards could be difficult. The prices offered by middlemen were on average 10 percent below the guaranteed prices, but they did provide transportation from farm to market (Price and Hall 1983). If the maize surplus could not be sold, it was usually lost because it does not store well. For example, the Choles in Frontera Corozal lost 400 tons of maize in 1978 because they could not transport it to the market (Mauricio Leguizamo et al. 1985).

Another crop raised primarily for subsistence is black beans. As the principal source of protein, many farmers cultivate them in intercrops. Nevertheless, beans are difficult to grow in the tropical environment, as they are highly susceptible to drainage problems as well as insect and disease damage. In fact, the Selva Lacandona region is a net importer of beans, with little of the local production sold in the marketplace (Price and Hall 1983). Bananas and plantains are also widely grown for home consumption, as are many other fruits and vegetables. Few are marketed, as prices are low in comparison to transportation costs. The products that are grown for the market include coffee, chile, and cacao, as well as illegal crops such as marijuana.

Coffee cultivation has increased dramatically in the Selva Lacandona over the past two decades, and has become the principal cash crop in many areas. Coffee is primarily produced at higher elevations in the western part of the region, between 500 and 1000 meters. It is considered to be a more viable and permanent system than crop or cattle production, as the shade canopy and coffee shrubs protect the soil. However, it is also a labor intensive activity, with the entire process involving 40 to 50 days of work for each hectare (Price and Hall 1983). Most farmers do not initiate coffee production as a first crop. Coffee bushes do not produce berries for three to six years, and recent colonizers cannot afford to devote labor to an enterprise which does not contribute to immediate subsistence (Price and Hall 1983). The influx of Guatemalan refugees in the early 1980s provided a source of cheap labor that allowed many recent colonizers to establish coffee production on their plots (Paz Salinas 1989).

The yields from coffee production in the Selva Lacandona vary significantly, depending on the variety planted, the quality of the land, and the production system employed. The coffee harvests are transported to the market in Comitán by various means, including chartered airplane. More frequently, however, the coffee is sold through intermediaries.

The importance of coffee production increased in many parts of the Selva Lacandona region during the 1980s because the financial returns were far better than for any other crop. By 1994, there were almost 17,000 producers in the Selva region, with 93 percent of them farming coffee on plots of less than two hectares (Harvey 1995). The state agency for coffee production, the *Instituto Nacional Mexicano del Café* (INMECAFE), played a central role in organizing and financing coffee production in other areas of Chiapas, as well as guaranteeing the purchase and export of the harvest (Harvey 1995). However, it appears that INMECAFE's role in the Selva Lacandona was limited to promotional activities. As early as 1974, it was encouraging coffee cultivation in some of the new ejidos by providing coffee bushes to those who wanted them (Preciado Llamas 1976). Nevertheless, in the 1980s, INMECAFE's role was insignificant:

> The Selva Lacandona has not benefited from INMECAFE. Officials of the Institute describe the Selva as an "intransitable" area, and a marginal producer. No services are provided to the region's producers. INMECAFE has no personnel in the region. There is also no government purchasing of coffee in the Selva (Price and Hall 1983:73).

Another crop which is produced solely for commercialization is chile. Chile production can be very profitable, if all goes well in the cultivation process. Chile production involves the use of inputs such as herbicides and insecticides, and is labor intensive. In 1979, the Selva Lacandona was the largest producer of dried chiles in Mexico. This can be attributed to low labor costs, which are applied to the labor-intensive drying process (Price and Hall 1983). Extension programs sponsored by the agricultural ministry promoted chile production among farmers in the Selva during the early 1990s. Many farmers planted chiles, only to find that they did not thrive. In addition, the market price for chiles is highly volatile, and there are no guaranteed government prices. Consequently, decisions to produce chiles involves substantial risk.

A third crop which is raised for commercial purposes in the Selva Lacandona is cacao. Cacao production, introduced to the region in the early 1980s, is intercropped with coffee in the some parts of the region and produced in pure stands in others. The government initiated a major cacao project in the Marqués de Comillas area in the early 1980s, creating extensive plantations. By 1986, cacao plantations covered over 2000 hectares. Production began four or five years after the trees were planted, and full production was reached after eight years (Price and Hall 1983). Like chile, cacao is a labor intensive crop prone to pest and disease problems. Consequently, only a fraction of the area planted could be harvested. By the early 1990s, most ejidatarios had, in fact, abandoned their cacao plantations because harvests were so low (Neubauer 1997).

While the physical environment of a tropical ecosystem can in some cases be a formidable barrier to production, it is just as often the case that economic constraints hinder sustainable agriculture. The migrants in the Selva Lacandona are tied to national and international agricultural markets and are vulnerable to their fluctuations. For example, when international coffee prices fell in 1989, many small-scale producers in the Selva Lacandona were adversely affected. Furthermore, the production inputs necessary to meet quality standards are largely unavailable to campesinos, and marketing mechanisms are highly underdeveloped. Credit, for example, is extremely difficult to acquire. In 1990, only 12.7 percent of producers in Chiapas held credit for agricultural production, with the percentage even lower in the Selva Lacandona region (Harvey 1996). In short, Price and Hall (1983:88) found the potential of agriculture in the Selva Lacandona to be high, but noted that there were many obstacles to overcome:

> Farming systems have been developed and adopted by farmers in the Selva Lacandona that are highly adapted to local environmental and economic conditions. In the early years of settlement, migrants are primarily subsistence farmers. Over time, mixes of subsistence and commercial enterprises are adopted. Multi-product systems are designed to maximize income, subject to the maintenance of subsistence, and subject to the availability of resources. . . . these systems have potential for relatively high levels of productivity and income generation. Many problems with both production and marketing have, however, resulted in the limited realization of this potential.

The failure of agricultural systems to provide a secure income for farmers in the Selva Lacandona has led many of them to turn to cattle production for cash income (Price and Hall 1983; Leyva Solano and Ascencio Franco 1996). Although the long-term effects on the ecosystem are devastating, the decision to raise cattle makes short-term economic sense for campesinos. The expansion of cattle in the Selva region is not an isolated example of irrational land use decisions, but instead forms part of a larger trend taking place throughout the tropics.

The Logic of Cattle Ranching

Cattle ranching merits special consideration because its impacts on the Selva Lacandona have been threefold. First, the expansion of cattle ranching throughout Chiapas has displaced agriculturalists and wage laborers, adding to the flow of migrants into the Selva Lacandona. Second, private cattle ranchers have been encroaching on land claimed and often cleared by peasants, creating conflicts that have led to protracted violence. Third, many ejidatarios and small-scale farmers within the Selva Lacandona have turned to cattle production as a supplement or alternative to crop production. In combination, these three forces have led to the transformation of extensive parts of the Selva Lacandona from forest to pasture.

Within Chiapas, the number of cattle has increased dramatically over the past six decades (Table 6.2). Nevertheless, cattle ranching is not a new activity to the region. Cattle were introduced by the Spanish during the colonial period. The early history of cattle ranching in Chiapas is related to the rise of the hacienda and the expansion of a mercantile economy (see García de León 1985; Villafuerte Solís et al. 1993). In Chiapas, cattle production was initially concentrated in the Central Valley and along the Pacific coast. During the Porfiriato, liberal land acquisition laws enabled cattle ranchers to expand their herds into the humid tropics of Chiapas. The number of cattle in the northern part of the state grew, as it did on the periphery of the Selva Lacandona, around Palenque and Ocosingo.

Prior to the Mexican Revolution, cattle from Chiapas were exported primarily to Central America, as well as to Yucatán and Tabasco (Fernández Ortiz et al. 1994). Cattle grazing did not become a significant form of land use until the agrarian reforms of Cárdenas were enacted in the 1930s. The minimum estate size for land dedicated to cattle production was much larger than for agricultural lands, and numerous laws protected these lands from expropriation (Reyes Ramos 1992). Cattle production thus offered landowners a means to avoid imminent land reforms. An alliance developed between cattle ranchers

TABLE 6.2   The expansion of cattle production in Chiapas, 1930-1990.

Year	Number of Cattle
1930	362,000
1940	423,000
1950	480,000
1960	682,000
1970	2,000,000
1980	2,935,000
1990	2,063,000

Source: Fernández Ortiz et al., 1994

and politicians in Chiapas that secured the interests of the landowning class. As a result, agricultural lands were increasingly converted to pastures and remained concentrated in the hands of the traditional *familia chiapaneca*. The production of cattle itself was not of great concern — landowners were largely interested in maintaining their extensive estates (Retiere 1991).

Cattle ranching as a production activity took off after 1950, as it did in many other tropical areas. Cattle replaced agricultural and plantation crops, both because the market was considerably more stable, and because there was great potential to develop a market within Mexico (Fernández Ortiz et al. 1994). Other reasons for the rapid increases in cattle ranching in Chiapas include the drop in international coffee prices, increasing national and international demand for meat, increases in credits for cattle production, and the legal protection provided to cattle lands through exemption certificates (Reyes Ramos 1992).

Between 1940 and 1980, successive state and national government policies and programs provided a host of supports and incentives to cattle owners. These included various means of expanding and consolidating landholdings, legal and fiscal protection of grazing lands, subsidies, development projects, and preferential credits (see Fernández Ortiz et al. 1994). The state government also encouraged cattle owners to organize themselves into associations and establish private police forces. Cattlemen's Associations serve commercial, social, and political functions in Mexico, including the provision of assistance to ranchers in the production and marketing of cattle (Price and Hall 1983). Two of the earliest associations were established in 1934 on the outskirts of the Selva Lacandona, in Ocosingo and Comitán.

As agricultural lands throughout Chiapas were converted to pasture, the campesinos who worked the land lost their jobs, adding to the unemployed rural population. Cattle production in Chiapas can be characterized as extensive rather than intensive. Aside from the initial investment in the animals, cattle ranching does not require a lot of capital and labor. Access to land, however, is essential in that each head of cattle requires an average of at least two hectares of pasture (García de León 1985). Price and Hall (1983) noted a case where a commercial cattle ranch with 750 hectares of pasture and 600 head of cattle provided employment for six full-time workers, 25 part-time workers, and one manager. Alternately, the land from this type of ranch could provide a livelihood for many families if it were held as an ejido. Instead of intensifying current land use and management practices to increase the stocking rate per hectare, commercial ranchers use small-holders to expand their operations (Price and Hall 1983).

Commercial cattle ranches hold advantages over small-scale ranches in terms of financing and marketing. More sophisticated animal husbandry practices also contribute to increased profitability. Although some small ranches produce finished cattle, most produce feeder cattle which are sold to larger ranches for fattening (Price and Hall 1983). In fact, ranchers from many parts of southeastern Mexico depend on small-holders from the Selva Lacandona to supply feeder cattle (Price and Hall 1983).

Several large, private ranches were established in the Selva Lacandona during the 1940s and 1950s, when it was still possible for individuals to purchase large landholdings or claim national lands. Price and Hall (1983) identified three ranches in the Selva that were believed to be over 1,000 hectares. They speculated that besides controlling large areas of land, these ranches also monopolized some of the best land in terms of agricultural quality. These three ranches have had an important impact on the growth of cattle production throughout the region.

With the exception of the few large ranches established within the Selva Lacandona, most of the early expansion in Chiapas was concentrated on former agricultural lands. After 1970, cattle production began to encroach into forested lands of the Selva Lacandona. Starting in the periphery, large landowners pushed deeper and deeper into the forest. In some cases, ranchers claimed land that had already been cleared by campesinos. In other cases, they rented pastures directly from ejidatarios or small private ranches. According to Leyva Solano and Ascencio Franco (1993), landowners allowed the campesinos to deforest parcels of land and cultivate it for several years, with the understanding that they would subsequently plant grass. Afterwards, they would be given a new plot to clear, and the process would continue until eventually the entire estate was a cattle ranch, with no room for agriculture. The campesinos would then be expelled from the property.

Conflicts over access to land have become commonplace in Chiapas, and the majority of them are between campesinos and cattle ranchers (Pólito and González Esponda 1996). In fact, of 117 conflicts registered between 1975 and 1977, 72 percent involved cattle ranchers (Fernández Ortiz et al. 1994). Many of the conflicts are over the forced removal of campesinos from their land by cattlemen or their private police forces. The lands in dispute include NCPEs, ejidos, communal lands and national lands.

After 1970, cattle expansion was increasingly focused on ejidal lands, and no longer limited to private ranches (Villafuerte Solís et al. 1993). By the 1980s, the ejidal sector in the Selva Lacandona was considered to be the only area that offered significant potential for further expansion of the cattle industry (Price and Hall 1983). Ejidos held the largest amount of land in the area, and all new colonization took the form of ejidos. Price and Hall (1983) identified several types of enterprises within the ejido sector. The two most common included pasture rental and the production of feeder cattle for fattening operations.

Pasture rental is characterized by the planting of grasses on cleared, fenced-in land, often by farmers who plan to purchase their own cattle in the future. Pasture may be rented to private ranchers or to other ejidatarios. Small-scale cattle production, on the other hand, starts out as an activity secondary to agriculture. Initially, only a few head of cattle are grazed on land that is considered to be the least productive for crops, such as low-lying land which is subject to flooding or water logging. The rate of cattle expansion depends on the availability of capital (Price and Hall 1983). As the area in pasture expands, it competes with agriculture for the most productive plots of land (Price and Hall 1983). In some cases, poor agricultural practices result in land which cannot be used for crops, and pasture becomes the only viable economic alternative (Price and Hall 1983).

Cattle production is adopted by small-holders for a number of reasons. First and foremost, it provides a higher income for farmers in a short period of time, relative to almost all other options available (Price and Hall 1983). By producing beyond levels of subsistence, families can afford clothes, medicine, improved housing, and other purchases. At the community level, profits from the sale of cattle can be put toward public works projects, travel costs to take care of agrarian issues in the state capital, and other expenses (Leyva Solano and Ascensio Franco 1993). Cattle can also be considered a means of storing wealth, enabling farmers to sell when prices are high or cash is needed. In contrast, most food crops must be sold immediately so that they do not spoil in the humid tropical environment.

Second, cattle production requires less labor than agriculture, and can initially be undertaken as a subsidiary activity (Leyva Solano and Ascensio Franco 1993). Unlike agriculture, it does not have a distinct cyclical nature, therefore labor can be contracted among friends and neighbors during the periods when crops require the least amount of attention (Lobato 1980). Finally, the availability of credits from the World Bank, INI, and the *Fideicomiso Instituido en Relacion con*

Agricultura (FIRA) in Mexico encouraged the expansion of cattle herds in the Selva Lacandona. Although a "cattle culture" existed since the onset of colonization (Leyva Solano and Ascensio Franco 1993), the credits enabled many communities to borrow the necessary capital to get started. Cattle expansion has been most rapid in areas of the Selva where credit has been available, and least rapid where it has been absent (Price and Hall 1983).

Cattle ranching is clearly an attractive alternative to agriculture for campesinos in the Selva Lacandona (Villafuerte Solís et al. 1993). By 1990, there were over 150,000 heads of cattle in the Selva Lacandona (Table 6.3). However, cattle ranching is not evenly distributed throughout the region. In some areas, the majority of campesino families raise cattle, and in others only a minority are involved in cattle production (Leyva Solano and Ascensio Franco 1993). Within the Comunidad Lacandona, there are over 6,500 hectares of pasture supporting 3,010 head of cattle (Villafuerte Solís et al. 1993). Nevertheless, the presence of pasture does not always reflect the number of cattle held by a family or community. In the Marqués de Comillas region, for example, campesinos quickly realized that land dedicated to pasture increased the security of eventual entitlement. Consequently, plots were deforested and used for only one cycle of maize cultivation before grasses were planted, despite the lack of cows (Mauricio Leguizamo et al. 1985). Pasture formation was also accelerated in anticipation of credits to purchase cattle. When the credits dried up in the late 1980s, many farmers in the Selva were left with empty pastures.

TABLE 6.3 Estimated number of cattle in the Selva Lacandona , 1990.

Municipio	Number of Cattle
Altamirano	17,083
Las Margaritas	33,358
Ocosingo	102,106
TOTAL	152,547

Source: INEGI, 1991.

The environmental consequences of cattle on the forest ecosystem are also variable. On the one hand, the impacts of those farmers with one or two heads of cattle can be considered trivial: "Farmers who have small sideline cattle operations do not have a strong impact, as a group, on the region's economy or ecology. Only the least productive land is generally planted to pasture" (Price and Hall 1983:101). On the other hand, "[e]xtensive cattle production, as it is practiced by most of the small-holders in the region is a nonsustainable system. These systems remain profitable only for a few years" (Price and Hall 1983:90).

The key factor in sustainable cattle grazing lies in pasture management, which includes rotational use of pastures and weed control. Unfortunately, many small ranches try to compensate for a lack of land by overstocking pastures, which is eventually destructive. Even a well-managed pasture is only productive for an average of six or seven years. After that, a fallow is necessary to replenish nutrients. If pastures are grazed for too long, soil degradation renders them useless, making it increasingly unlikely that tropical forests can be successfully reestablished. Within some parts of the Selva Lacandona, cattle ranching has displaced agriculture from flat lands to hillsides, leading to greater soil erosion and diminished yields.

The cattle sector in Mexico contracted dramatically after the economic crisis of 1982. According to official statistics, the total herd size fell from 37.5 million heads in 1983 to 31.5 million in 1991 (Villafuerte Solís et al. 1993). Nationally, public investment in the sector decreased during the 1980s, and credits dried up. Within Chiapas, the 1980s were marked by erratic fluctuations in the size of the cattle herd. Nevertheless, the Selva Lacandona region, including Palenque, stands out as the most dynamic area in terms of cattle production. Despite a lack of credits, cattle ranching has become an integral part of the internal regional economy. In fact, from a strategic perspective, it is considered a viable and often essential activity for survival (Villafuerte Solís et al. 1993).

Living on the Edge

Colonization of the Selva Lacandona has opened up one of the last agricultural frontiers in Chiapas. However, the notion of an agricultural frontier misleadingly overstates its role in the national production system. In this case, the expansion of the area under production was not strategically intended as part of a national program to increase agricultural production. Instead, it served to maintain the large estates elsewhere in Chiapas that were seen as engines of economic growth. According to the analysis of Preciado Llamas (1978), the development of capitalist agriculture within Mexico was actually impeded in the short term, in that colonization favored small-scale farmers producing for self-sufficiency. In the medium to long term, however, colonization did open up new territories for the penetration of capital. By the medium term, Preciado Llamas refers to the time it takes for social differentiation to develop among the colonizers, and for the state to construct sufficient infrastructure for capitalists to penetrate the newly opened spaces at little cost and high profits (Preciado Llamas 1978).

Social stratification has developed within and among the communities of the Selva Lacandona (Lobato 1980; Lobato 1981). In particular, there is a clear distinction between the communities in the subregions of Las Margaritas and Las Cañadas, where colonization was more often induced by agrarian policies, than in the Marqués de Comillas and Comunidad Lacandona subregions, where colonization was to a greater degree affected by government direction or intervention (Leyva Solano and Ascencio Franco 1996). Yet while some entrepreneurial ejidatarios, small ranchers, or individuals have been able to prosper in the area, the majority of people in the Selva Lacandona are living on the edge, trying to eke out a living in a challenging physical environment. The production strategies of most residents of the Selva Lacandona are targeted beyond subsistence, in hopes of producing a surplus, acquiring capital, and securing a better future for their children.

7

Refugees and Rebels

Political Upheavals

The previous chapters have described how the Selva Lacandona has served as a source of rapid capital accumulation, a focus of land speculation, a refuge for displaced and disempowered people, and a political safety valve for concessionary politics. The region has also been affected by political upheavals, which have led to an influx of refugees as well as the militarization of parts of the region. These upheavals have had an impact on the region´s social and production relations, as well as on its ecology.

The first upheaval was related to the guerrilla war in Guatemala, which sent thousands of peasants across the political border into Mexico between 1980 and 1985. The presence of these refugees increased the population in the forest as well as the amount of the land under cultivation, particularly in Las Margaritas and the Marqués de Comillas region of Ocosingo. Moreover, the availability of cheap labor allowed many Mexican campesinos to expand production of market-oriented crops. Although the presence of refugees has declined as a result of the United Nations-sponsored repatriation efforts initiated in 1993, a number of Guatemalan refugees continue to live in the Selva Lacandona.

The most recent political upheaval to affect the Selva Lacandona has been the 1994 rebellion by the *Ejército Zapatista de Liberación Nacional* (EZLN). This military uprising and political rebellion is closely linked to the history of agricultural transformations and the resulting agrarian crisis in Chiapas. The uprising brought not only international attention to the "jungle of Chiapas," but also a heavy military presence. The conflict is based in the western part of the forest, in the municipios of Ocosingo and Las Margaritas, particularly in the subregion of Las Cañadas.

Tropical forests have been both a direct and an indirect casualty in these conflicts. In this chapter, the causes and consequences of the influx of refugees from Guatemala are considered. The roots of the Zapatista conflict are also discussed, along with the implications for the remaining forest. These two examples illustrate how deforestation in the Selva Lacandona region has been influenced by both external and internal political upheavals. The tensions between environmental and social struggles in the Selva Lacandona have placed the remaining forest at the center of small but growing conflicts.

Guatemalan Refugees

The Selva Lacandona shares extensive borders with Guatemala, including the Usumacinta and Salinas/Chixoy rivers to the east, and an east-west border at 16°04'20" N. latitude. While border disputes in these areas created tensions between Mexico and Guatemala in the nineteenth century (de Vos 1988a), the issue of refugees has dominated the relationship in the late twentieth century.

Guatemala's long history of political violence, military repression, and human rights abuses was underscored in the 1970s and early 1980s with the massacres of thousands of peasant farmers living in the Guatemalan highlands and in the lowland rain forests of the north. Following these massacres and the militarization of the countryside, an estimated two hundred thousand refugees fled to Mexico and the United States (Manz 1988). Many of the refugees escaped across the border into the Selva Lacandona, which is an extension of the lowland rain forest of Guatemala. Approximately 46,000 Guatemalans set up camps within communities in the region, and many remained despite resettlement programs organized by the Mexican government and the United Nations.

The overt cause of Guatemala's war in the countryside can be attributed to the presence of guerrilla armies supported by the peasantry. The emergence of radical opposition forces can be traced to the repressive policies of successive governments aiming to control land and labor throughout the country. The overthrow of democratically elected President Jacobo Arbenz in 1954 led to twelve years of military rule in Guatemala (Castellanos Cambranes 1984). Repression varied during those years, and was in fact exacerbated when a civilian government came to power in 1966. The repression did not abate until 1974, when General Kjell Eugenio Laugerud García became president. The relaxed political environment benefitted grassroots organizations, some of which had managed to survive the particularly repressive years of 1966 to 1974 (Manz 1988). Without directly challenging the government, these grassroots movements flourished and were able to contribute to the improved living conditions of its members. Indigenous groups of the highlands also entered into the national political scene for the first time (Stepputat 1989).

Although the overt cause of the flow of refugees into the Selva Lacandona was the war between the Guatemalan army and guerrilla forces, the underlying causes can be traced to access to land. While a thorough analysis will not be presented here, some similarities with the situation in Mexico are striking and worth mentioning. Guatemala possesses some of the most fertile agricultural land in Central America, yet has one of the most skewed land distributions in Latin America. Over half of the country's population of 8.7 million earns a living from agriculture, yet more than one-third of the farmland is included in 1 percent of all farms. Over half of all farms are smaller than 1.5 hectares, and they occupy only 4 percent of the land. What is even more surprising is that over 1.2 million hectares of arable land, the vast majority of it in farms of more than 450 hectares, is lying idle or is underutilized as grazing land. Meanwhile, the land needed for landless peasants amounts to approximately the same (Manz 1988).

While a large amount of land was underutilized in Guatemala, approximately 400,000 families, many of them from indigenous groups in the highlands, spent a great deal of time wandering around in search of land. With the exception of a land reform passed by the Arbenz government prior to its overthrow in 1954, land redistributions were simply not an issue in Guatemala. What the post-Arbenz governments opted for instead was the resettlement of thousands of landless families on undeveloped public lands in the Petén, and in the lowland rain forests in the northern parts of Huehuetenango, El Quiché, and Alta Verapaz. This was organized through the Department of Colonization and Agricultural Development, and involved the colonization of national lands, as in the case of the Selva Lacandona.

Colonization was also organized by religious leaders in the late 1960s and early 1970s. The sparsely inhabited area around the Río Ixcán was first colonized in 1966 by a Maryknoll priest and 14 Indian followers, who were soon joined by a group of Mam Indians (Manz 1988). The area east of the Río Chajul/Xalbal was settled by a Spanish priest and Quiche Indians in 1972. The latter group had been successfully organized into production cooperatives in the highlands, but migrated in search of land. According to Manz (1988), the peasants realized that insufficient land lay at the core of their problems, and given the unyielding opposition to land reform by the Guatemalan government, the only hope appeared to be colonizing the tropical forest to the north.

Eventually, grassroots organizations reached a limit to what they could accomplish by working within the system. The military and economic elites viewed any reform as revolutionary and responded with fierce opposition (Manz 1988). As a result of this political inflexibility, individuals and entire communities that had been searching only for reform became radicalized.

In addition to the formation of nonviolent grassroots movements during the 1960s, guerrilla forces emerged in opposition to the ruling elite of Guatemala. The *Fuerzas Armadas Rebeldes* (FAR), formed in 1962, operated primarily in the ladino areas of the country, such as the eastern coast and Guatemala City. This guerrilla force was destroyed in 1966 by an army attack that killed 9,000 civilians as well. Although short-lived, the FAR prepared the ground for future guerrilla movements, including the *Organización Revolucionario del Pueblo en Armas* (ORPA) and the *Ejército Guerrillero de los Pobres* (EGP) (Falla 1994).

The EGP began to operate politically in the Ixcán region of Guatemala during the 1970s. Although some peasants actively participated in the guerrilla movement and many sympathized with the cause, the majority of the colonizers in the lowland rain forest were more interested in making a living from the land. The army's policy, however, was to eradicate the guerrillas by killing all potential supporters, which included most civilians. Beginning in 1975, the army carried out "disappearances, torture, selective killings, executions of several people at a time, group massacres, and massive massacres of entire villages — in a word, genocide" (Falla 1994:4). The scorched earth policy depopulated much of the area and eventually gave rise to *Comunidades en Resistencia* (communities in resistance), made up of people who lived in hiding in the forest, outside of army controlled territory. The massacres continued into the mid-1980s with the torture and murder of men, women and children in villages in the tropical forests of Guatemala. These massacres resulted in an outpouring of refugees across the border into Mexico.

According to Manz (1988:146), "[t]he first large-scale arrival of Guatemalan refugees took place in May and June 1981. These people came from cooperatives in El Petén founded in the 1960s along the Usumacinta and La Pasión rivers, which were attacked by soldiers from the military base at Poptún. Villagers sought refuge in Chiapas, and found warm hospitality." After the 1982 military counterinsurgency drive in the Ixcán region of Guatemala, the flow across the border surged (Manz 1988). Tens of thousands of refugees, malnourished and diseased, entered Mexico after the Guatemalan army carried out massacres in hundreds of villages. By December 1982, there were 56 camps with 36,000 Guatemalan refugees spread along the border from Ciudad Cuauhtémoc to the edge of the Lacantún River in the Marqués de Comillas region of the Selva Lacandona. By 1983, the number had risen to 46,000 refugees in approximately 90 camps (Manz 1988). The main camps in Chiapas and their refugee populations are identified in Figure 7.1. The largest camps were found in the Selva Lacandona: Puerto Rico had a population of 5,080, Boca de Chajul had 3,150, and Ixcán had 2,877. Manz (1988:147) explains that "[t]he large-scale exodus of entire villages and the absence of many Mexican communities in that inaccessible area accounts for the size of the camps and concentration of refugees."

By 1983, the military had established control over the region. The hundreds of farms and villages that were abandoned after the massacres were slowly repopulated by landless peasants imported by the military from other parts of the country (Manz 1988). The government established a rural development plan based on regional centers located in areas of strategic importance. These "development poles" physically concentrated dispersed settlements into larger and more easily controlled units (Manz 1988).

Meanwhile, in Mexico, federal and local officials were fearful that the influx of refugees would lead to political, economic, social, and military turmoil in Chiapas, including the possibility of a guerrilla war. The Mexican government addressed these fears by increasing its military presence in Chiapas, and by financing a road along the border to establish a Mexican presence in an otherwise remote area. When it became clear that the situation in Guatemala would not improve, the government initiated a large-scale relocation of the refugees to other states in southern Mexico. The relocation, which began in 1984, was resisted by many of the refugees. For the most part, the resistance centered around geographic and cultural concerns. Given that the border is a political division rather than a geographical or cultural divide, many of the refugees felt strong ties to their homes while in temporary exile in the Selva Lacandona (Manz 1988).

Although UN-sponsored repatriation talks got underway in 1992, by 1994 an estimated 23,000 refugees were still in Chiapas, including about 12,000 located in 72 camps in the Selva Lacandona (Castro Soto 1994). Repatriation efforts have been marred by continued violence in the Guatemalan countryside. Those who do return often find their original lands occupied by new settlers invited in by the military, and thus

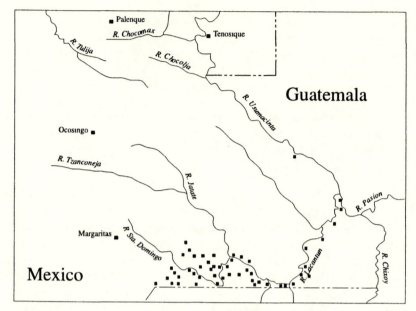

Camp	Refugees	Camp	Refugees
Alvaro Obregón	79	Monte Flor	1,928
Amparo Agua Tinta	1,258	Niños Heroes	258
Bella Illusión	36	Nuevo Huixtán	383
Benemérito de las Américas	678	Nuevo Jerusalén	251
Chajul	3,150	Nuevo Santo Tomás	333
Cieneguitas	1,520	Peña Blanca	131
Flor de Café	553	Pico de Oro	1,750
Frontera Corozal	470	Playón de la Gloria	201
Galaxia	62	Poza Rica	1,235
Gallo Giro	158	Puerto Rico	5,080
Guadalupe Miramar	81	Quiringuicharo	808
Ixcán	2,877	Rancho Alegre	114
José Castillo Tielemans	210	Reforma Agraria	363
La Constitución	212	Rizo de Oro	443
La Democracia	49	San Antonio los Montes	96
La Gloria de San Caralampio	2,582	San Carlos del Rio	76
Las Delicias	606	San Juan Chamula	501
Las Ventanas	840	Santa Margarita Agua Azul	124
Loma Bonita	259	Santo Domingo Las Palmas	148
López Mateos	282	Venustiano Carranza	270
Maravilla Tenejapa	149	Vicente Guerrero-Carman Xhan	268
Monte Cristo	93	Zaccualtipán	68
		TOTAL	31,033

FIGURE 7.1 Refugee camps in the Selva Lacandona, 1984 (*Source*: Manz, 1988).

have to reinitiate the colonization process on new lands, including the tropical forests of the Maya Biosphere Reserve in the Petén region.

The Zapatista Uprising

The most recent event which has brought the Selva Lacandona to the attention of the national and international communities has been the rebellion of the *Ejército Zapatista de Liberacion Nacional* (EZLN). This peasant uprising, based in the Selva Lacandona, captured headlines for months, and was responsible for several crises of confidence in the Mexican government. The conflict began on January 1, 1994, when hundreds of armed and unarmed Zapatista soldiers took control of a number of towns in Chiapas, including San Cristóbal de las Casas, Ocosingo, Altamirano, and Las Margaritas. The uprising coincided with the official beginning of the North American Free Trade Agreement (NAFTA), a fact which did not escape notice. The Zapatistas presented their demands to the media, the government, and the general public through the Declaration of the Lacandón Jungle: Today we say "Enough." The goals of the struggle were outlined in this document by the General Command of the EZLN in 1993:

> People of Mexico. We, men and women, upright and free, are conscious that the war we now declare is a last resort, but it also just. The dictatorship has been waging a non-declared genocidal war against our communities for many years. We now ask for your committed participation and support for this plan of the people of Mexico who struggle for work, land, housing, food, health, education, independence, liberty, democracy, justice and peace. We declare that we will not stop fighting until we win these basic demands of our people, forming a free and democratic government. 1

The emergence of a rebellious uprising took the country by surprise, despite the fact that rumors of its existence had been circulating for some time. In fact, Mexican authorities had confronted Zapatistas in June, 1993, but the encounter was played down because the government did not want to endanger NAFTA or scare off potential investors. The January actions, in contrast, evoked a heavy military response. Air and land attacks began on January 3, and continued until a ceasefire was declared on January 13, 1994.

The uprising led to a flow of refugees out of villages in the Selva Lacandona and parts of the Highlands. Human rights abuses were reported for both sides, and a fear of an impending military escalation pervaded the country. In the subsequent weeks, civil society mobilized in demand of a peaceful solution to the conflict, and official peace negotiations began in San Cristóbal on February 21. After consulting with its support base, the EZLN rejected the resulting peace proposal. Since then, a tense ceasefire has been maintained. Moreover, the Zapatistas have formed a political movement that is challenging the neoliberal development model and working to instill a true democracy in Mexico. An integral part of the Zapatista vision includes full rights and a recognition of autonomy for indigenous groups. Negotiations with the government have been continuously delayed until the Zapatistas are convinced of the government´s commitment to real change, rather than to the rhetoric of appeasement.

Meanwhile, the Mexican army has permeated the Selva Lacandona, conducting what many feel to be low-intensity warfare (Lopez A. 1996). The military´s expanded presence in the region can be seen in Figure 7.2. The presence of over 30,000 soldiers has disrupted life in many communities in the Selva Lacandona. In a number of villages, the army outnumbers the local population. Prostitution, alcoholism, and inflationary prices for goods have pervaded some of the communities where the army maintains a strong presence (Balboa 1997).

The Zapatista Uprising is often said to have arisen in the heart of the remote Lacandón jungle. While the area in which the rebels are based can be considered remote and difficult to access, in reality the area has been extensively cleared through the process of colonization. While many of the mountains still remain forested and impede easy access to the area, the valleys consist of a mosaic of agricultural lands. While the Zapatista-controlled areas do fall within the Selva Lacandona region, it would be a mistake to ascribe the rebellion to all communities within the Selva, or even to all individuals within a community. The uprising is largely restricted to the western part of the forest, in the mountains and valleys to the southeast of Ocosingo and to the east of Las Margaritas. The Comunidad Lacandona did not participate in the uprising, nor did the settlers in the Marqués de Comillas region.

Ostensibly, the Zapatista Rebellion has been considered an indigenous-based uprising caused by extreme poverty and marginalization. Although many people frame it as an exclusively Indian rebellion, others feel that it is more accurately described as a peasant rebellion (Collier and Quaratiello 1994). Some have alleged that land degradation was responsible for the uprising, implying that irrational land use practices led to the marginalization that fostered the rebellion. For example, Carlos Rojas Gutiérrez, the head of Mexico's social development agency, the Secretaría de Desarrollo Social (Sedeso), contended that the social problems in the municipios located

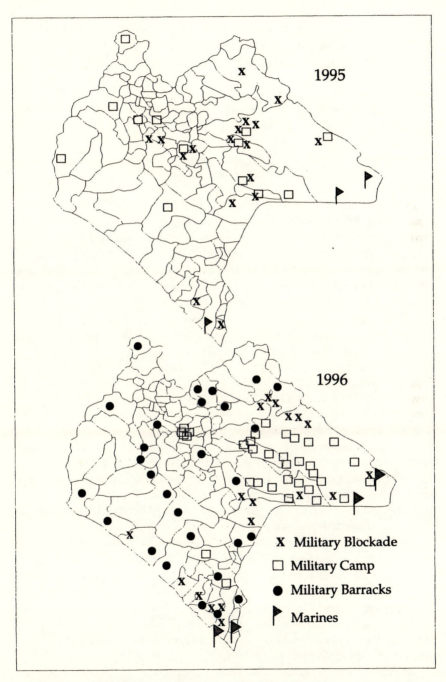

FIGURE 7.2 Militarization of Chiapas, 1995-1996 (Adapted from Conpaz et al. 1996).

in the Selva Lacandona worsened due to demographic growth and the deterioration of natural resources (Elvira Vargas 1994). Similarly, Howard and Homer-Dixon (1996) attribute the uprising to growing environmental scarcities, including demand-induced scarcity, supply-induced scarcity, and structural scarcity. In reality, the Zapatista Uprising is closely tied to the same agrarian politics that led to the colonization of the Selva Lacandona.

Peasant Mobilization

A latent conflict has been kindling in Chiapas for the past century, punctuated by periodic outbreaks of violence. Throughout the 19th century, conflicts in Chiapas revolved around the Liberals of the Central Valley and the Conservatives of the Highlands (see Benjamin 1989). Only occasionally did the threat of indigenous uprisings interrupt these conflicts and temporarily divert attention from regional rivalries. One such case occurred between 1867 and 1870, when indigenous groups in the Highlands sought political autonomy from the ruling elite of San Cristóbal. The Caste War was inspired by a supernatural experience of a Tzotzil Indian, and it led to the siege of San Cristóbal by indigenous rebels. Similar to an earlier revolt in the 18th century, this one was accompanied by a shift in the balance of power away from the ecclesiastical and toward secular powers (Marion Singer 1988). The war became an extremely violent interethnic conflict that left much of the state in ruins for a number of years. A total of 800 indigenous peasants were killed, as well as 200 ladinos. The uprising bred such a fear of recurrence that for many years the elite of San Cristóbal would only have to suggest a "caste war" to obtain state and federal intervention in agrarian issues (García de León 1984a).

Generations after the indigenous groups of the Highlands rose up against the landed elite of Chiapas in 1867, the flame of insurrection was still being kindled by peasants in opposition to landowners (Marion Singer 1988). According to Marion Singer (1988:49), the latent antagonism can be considered the historical legacy of the secular conflict which placed the landless wage laborers in conflict with their landowning bosses. Marion Singer further notes that the solution to this conflict was not achieved through the war of 1869, nor through the Mexican Revolution, nor through the reform politics and agricultural modernization of the twentieth century.

The migration of thousands of campesinos into the Selva Lacandona during the 1960s did not resolve agrarian problems in Chiapas either.

Although the ejidal sector expanded in terms of landholdings, ejidos throughout Chiapas were overpopulated by the 1960s (Benjamin 1989). In many cases, ejidal plots were too small to support a family, and ejidatarios were forced to work on neighboring estates, migrate for seasonal employment, or uproot their families and seek land in the Selva Lacandona. By the 1970s, the tensions over access and titles to land were beginning to take on violent characteristics (Pólito and González Esponda 1996).

The growing pressure for land coincided with a period of economic crisis in Chiapas. The economic crisis that began in the early 1970s was related in part to the integration of the state into national agricultural development plans of President Luis Echeverría (Benjamin 1989). The crisis was exacerbated by a growing population and an influx of Central American refugees to Chiapas. Campesinos were also displaced from the land through the expansion of cattle ranching, the construction of three hydroelectric dams, and the volcanic eruption of El Chichonal in 1982.[2] Together, these factors resulted in an increased questioning of the legitimacy of latifundios, and the emergence and radicalization of campesino movements.

In the 1970s, agricultural wage laborers in Chiapas began to denounce the archaic structures of exploitation that "maintained them in a state of virtual economic, political, and sociocultural domination" (Marion Singer 1988:40, own translation). These denunciations were largely a result of the expanding influence of independent campesino movements. The growth of these movements and their role in the Zapatista Uprising is described in detail by Harvey (1995).

Actions followed the denouncements, including land takeovers, marches, and demonstrations. However, because the actions were largely unorganized or isolated incidents, state and federal authorities were able to use repression to preserve peace and order and protect private property (Benjamin 1989). Rather than acknowledge the emergence of a powerful agrarian movement, officials made the activists out to be bandits, thieves, or drug traffickers.

Meanwhile, large estates continued to be fortified against potential breakups or takeovers, both by national and state policies. The use of exemption certificates to protect landholdings increased. Their use had been rare prior to the 1970s, when national lands were available to satisfy the growing demand for land. However, national lands were running out, and large estates were facing increasing pressure. The state government promoted exemption certificates in an unprecedented form. During the administration of governor Absalón Castellanos Dominguez (1982-1986), 2,932 agricultural certificates were issued, covering an area of 52,742 hectares (Reyes Ramos 1992). Over 4,000 livestock certificates were also issued, protecting 1.1 million hectares of ranch land (Reyes Ramos 1992).

At the political level, the state government considered agrarian reform to be finished by the late 1970s (Reyes Ramos 1992). The focus was instead directed toward increasing productivity on existing farms. As Reyes Ramos (1992:107) observed, the government continued to view the problem of land tenure as exclusively a production problem, rather than facing the inequities of land distributions. Proposed solutions to the agrarian crisis followed this mode of reasoning. For example, governor Juan Sabines Gutiérrez (1979-1982) created production contracts known as *Convenios de Confianza Agrario* with 1,800 ejidos (Reyes Ramos 1992). These contracts included a number of programs aimed at increasing production through increased investments in seeds, tractors, and other inputs. Governor Castellanos Domínguez later initiated a program for the creation and development of six districts of agricultural rehabilitation (Reyes Ramos 1992). The intention was to buy land from private owners to sell to the campesinos, creating small property owners rather than ejidatarios.

Despite the efforts of large landowners to legally protect their estates, land takeovers became a common occurrence in Chiapas during the 1970s and 1980s. In most cases, campesinos were not randomly invading land, but instead they took over land which had been solicited years earlier. This reflects what can be considered a source of permanent conflict in Mexico — the non-execution of presidential resolutions (Reyes Ramos 1992).[3]

A Region Revolts

Since the onset of the Zapatista Uprising, a tremendous amount of literature has been published recounting the events as they unfolded. These have included commentaries and analyses of the causes and context under which they occurred (Trejo Delarbre 1994; Romero Jacobo 1994; Monroy 1994; Collier and Quaratiello 1994; Katzenberger 1995; Russell 1995; Ross 1995). Most of the reports focus on the poverty and misery of the campesinos, the relationship to agrarian structures and the neoliberal economic policies of the past decade, and the emergence of social movements in the region. In particular, changes to Article 27 of the Mexican Constitution and the passage of the North American Free Trade Agreement (NAFTA) are considered to have stimulated unrest in the countryside. The links to poverty, land tenure issues, macroeconomic policies, and peasant mobilization are indeed considerable. However, these links are common to other parts of Chiapas and Mexico, where an armed uprising did not result. Why, then, did the rebellion emanate from the Selva Lacandona?

In many ways it is not surprising that the rebellion emerged from the Selva Lacandona, rather than from the many other marginalized areas in the state or country. The area is remote and poorly connected with the social and economic centers of Chiapas. Medical facilities and educational opportunities are notably lacking. The poverty and isolation of many of the communities make them ideal candidates for organization, as catechists from the Catholic Church have learned (Floyd 1994). The injustices committed against peasant communities and agrarian activists by landowners, backed by the state government, have also made the area ripe for political mobilization (Harvey 1995).

Leyva Solano and Ascencio Franco (1996:180) attribute the fact that the Marqués de Comillas and Comunidad Lacandona subregions did not participate in the revolt to the unequal amount of governmental attention paid to these areas. Since the mid-1970s, the two subregions have been the focus of a disproportionate amount of resources, including agricultural production projects, roads, clinics and housing projects. In comparison, regions such as Las Cañadas were left to organize and mobilize in their struggle for land tenure, and for meager support from INMECAFE and CONASUPO.

It was argued in the previous chapter that opening the lands of the Selva Lacandona to colonization was not so much in the interest of expanding agricultural production, but rather in diverting the focus of agrarian reforms away from the breakup of large estates. As such, many of the factors necessary for market participation, such as transportation networks and marketing facilities, were not firmly established. This is a common characteristic of recently colonized areas. Despite government promises, colonizers in many parts of the forest were left to fend for themselves (Paz Salinas 1989). Settlers in the Marqués de Comillas region found Guatemala to be an accessible market until the early 1980s, when government military repression in Guatemala devastated the settlements. Most areas, however, remained isolated from regional markets. Price and Hall (1983:117) noted the absence of markets in their study of agricultural development in the Selva Lacandona region:

> In the Selva Lacandona today, there are essentially no local or regional market centers. The individual settlements are not strongly linked economically, and within the colonies there are no village markets. The lack of market centers reduces the possibility of locally initiated commercial activity.

In years when production surpassed subsistence needs and offered the possibility of semi-commercialization, campesinos were approached by intermediaries, or *coyotes*, who had access to transportation. This often resulted in an exploitative relationship. Manuel Lombera, founder of the ejido Boca de Chajul in the Marqués de Comillas subregion, recalls that one such middleman arrived and offered them 50 centavos per kilo of corn and one peso per kilo of beans. Recognizing that the man was taking advantage of them, they refused to sell, in spite of an urgent need for income (de Vos 1988b).

Colonizers were also subjected to exorbitant prices by intermediaries for food, clothing, medicine, and consumer goods. To avoid reliance on itinerant merchants, some communities would charter a small plane and purchase provisions in Comitán or other large markets. In other cases, entrepreneurial individuals from the communities would acquire provisions to sell in a local *tienda*. Food items could also be purchased for subsidized prices in government-run CONASUPO stores, which were located in some of the larger villages.

The demand for goods produced outside of the ejido meant that a cash income was essential. Even the smallest farmers would sell some of their harvest to neighbors or nearby communities for cash. However, many would travel to the markets of Comitán, Las Margaritas, Ocosingo, Palenque, or Guatemala to sell their surplus. These journeys were long and demanding, and the rewards were uncertain. International price fluctuations could render the crops worthless and leave farmers in debt, as was the case when international coffee prices fell in 1989.

In other parts of Chiapas, peasants faced with economic hardship were able to seek wage labor in other sectors of the economy. Collier et al. (1994:398) found that "peasants are flexible and respond rapidly to macroeconomic change, shifting the balance between varied farming practices and off-farm activities in relation to changing rewards for their labor and other resources." Although such flexibility does not apply equally to all peasants, proximity to a market for labor offers opportunities for generating an income when agricultural markets do not.

In the Selva Lacandona, off-farm income opportunities are notably absent. Occasional exceptions occur, such as temporary work for PEMEX during the early 1980s. In socially stratified communities, seasonal wage labor is sometimes available on the largest farms. In general, however, a diversified peasant economy has not emerged. Indeed, the development of secondary and tertiary sectors has been largely ignored in the Selva Lacandona region. The employment problem was foreseen over a decade ago, when Price and Hall (1983:120) carried out a study in the region:

> Development efforts in the Selva Lacandona have completely ignored the issue of unemployment and under-employment. . . . Many of the migrants interviewed looked for work off farm. Most did not find sufficient employment during the year. The scarcity of wage labor jobs in the Selva is a particularly serious problem for families with little or no land. Some of these landless migrants are almost entirely dependent on wage laboring for a livelihood. Many look for work outside of the region. The majority of ejidos in the Selva Lacandona have little or no land in reserve for the children of the original migrants. In the future, there will be growing numbers of landless families as a result.

Despite the remote location of the Selva Lacandona, most settlers are indeed dependent to some extent on the market for their subsistence. Subsidized foods, guaranteed prices and the availability of limited credits and marketing facilities formed part of the survival strategy. Although government agricultural policies were not ideal, they did provide a safety net for survival. The gradual shift toward liberal economic policies and integration of agriculture in the global economy has left many campesinos quite vulnerable. Given that the political system was closed to their cries, it is not surprising that an armed rebellion emerged.

Ecological Implications

Both of the political upheavals described above have had significant impacts on the forests of the Selva Lacandona. The impacts have been both direct and indirect, influenced by the movement of refugees into and out of the region; an increase in military activities; an expansion of government participation in the region; and changes in social and production relations resulting from the aforementioned factors.

The flow of thousands of Guatemalan refugees into the Selva Lacandona region had multiple effects. First, the influx of refugees meant that more forest was cleared to establish the refugee camps. Second, the refugees provided a cheap source of labor for Mexican campesinos and landowners to exploit, allowing them to clear more land to expand production. Finally, the refugee situation led to a stronger Mexican presence along the border, both military and civilian.

Three types of refugee camps were created in the Selva Lacandona region. These included settlements on ejidos, dispersed among the Mexican population; concentrated settlements located within ejidos;

and concentrated settlements located on private property (Hernández Castillo 1992). The most common pattern of settlement was the first type, where refugees lived on ejidos in dispersed dwellings, mimicking the conditions of their home villages in Guatemala (Hernández Castillo 1992).

In many cases, the population of ejidos increased from five to ten times its original size. Private landholdings that accepted refugees soon matched small or large communities in size. With each refugee family holding at least one hectare of land to cultivate corn for subsistence, significant amounts of forests were cleared to accommodate the temporary settlers. Trees were also cut to construct shelters. Eventually, all of the camps had stable and permanent constructions (Manz 1988). In addition, many refugees were able to expand agricultural production beyond subsistence. Manz (1988:147) describes the situation in refugee camps as follows:

> Once the refugees had met basic needs such as shelter, food and medicine, they organized the construction of the camps. In fact, the outstanding feature of the refugee camps in Chiapas was their self-organization. Shelters that initially consisted of a plastic sheet or branches became regular huts made of thatched roofs with walls of wooden poles. Soon the camps had pathways, central communal buildings, and a plaza with a kiosk. Even roads and airstrips were constructed in the virtually impenetrable Lacandón Forest. . . . The work was done communally. When hundreds of people arrived at a camp, for example, everyone would become involved cutting down trees to build a shelter.

Although the Mexican government sought to resettle the refugees in Campeche and Quintana Roo as early as 1984, many of the refugees chose to remain in the Selva Lacandona, for both geographic and cultural reasons (Manz 1988). The refugees that stayed on in the Selva Lacandona provided a cheap labor supply to Mexican campesinos, cattlemen and landowners (Nash 1995). In some cases, property owners paid less than the minimum daily wage to refugees, in exchange for land to cultivate, as well as for access to water and fuelwood (Hernández Castillo 1992). In other cases, ejidatarios or landowners were compensated for their land with money or payment in kind. As a result, many Mexicans were able to expand their production at rates that under normal circumstances would not have been possible. Refugee labor was used to perform the tedious task of cutting forest to prepare fields, and to harvest labor intensive crops such as coffee.

Finally, the refugee situation led to a stronger Mexican presence along the border. Military incursions across the border by the Guatemalan army resulted in heavier militarization of Chiapas, particularly in the border region. The government sponsored the construction of a highway along the Mexico-Guatemala border, as discussed in Chapter 4.[4] The presence of this road facilitated colonization of the rain forest, in this case as planned settlements organized by the government. Between 1980 and 1986, 10 ejidos were established along the border in the Marqués de Comillas subregion, covering an area of 49,521 hectares (González Ponciano 1990). Settlements in Las Margaritas subregion also benefitted from the rapid completion of the road from Comitán to Flor de Café. Once the road was finished, numerous institutions arrived to help the refugees, along with the Mexican colonizers who for had had for so long been ignored (Paz Salinas 1989). In short, the war in Guatemala provided an impetus for integrating some of the most remote parts of the Selva Lacandona with the rest of the state. This integration spurred another wave of deforestation.

As with the Guatemalan conflict, the Zapatista Uprising has had both direct and indirect implications for the ecology of the region. The conflict initially gave rise to a spate of rumors regarding military responses in the Selva Lacandona. These ranged from the intended use of Agent Orange to defoliate the forest to the poisoning of the rivers to kill off rural supporters of the Zapatistas. Fortunately, these rumors proved to be unfounded.

The actual consequences of the Zapatista Uprising can be considered threefold. First, an expanded military presence in the region has resulted in an amplification of deforestation pressures. An estimated 30,000 to 40,000 soldiers have established camps within the Selva Lacandona. Many argue that the Mexican Army is conducting low-intensity warfare under the guise of a humanitarian operation that seeks to militarize social services:

> The "Army for Peace," as it calls itself, initiated programs to hand out food, provide medical attention, and repair schools. . . . These programs, besides seeking to justify increasing military presence, serve to divide communities by providing aid selectively to campesinos who participate in government-allied organizatons and by harassing others (Carlsen 1996:3).

The establishment of military camps to support thousands of soldiers in the Selva has also had severe impacts on the social and economic relations within occupied communities, as well as on the local ecology. A fact-finding mission carried out by Conpaz et al. (1996) documented the impacts of militarization on a number of communities within the Selva Lacandona. They compiled an extensive list of actions, including the following impacts on the environment. The Army felled trees along the margins of the Río Jataté and Laguna Miramar; penetrated into remote mountains and the area around Laguna

Miramar, establishing paths and clearings; cut wood and timber for their constructions without authorization from local communities; hunted wildlife and promoted trafficking in wildlife; and deposited garbage into rivers and lakes. The social and economic impacts of militarization are even more extensive, including a scarcity of food and price inflation; reduced agricultural productivity through the restriction of free movement to plots; theft of food and animals from community members; increase in alcoholism; introduction of prostitution into communities; and increased polarization and distrust within communities. The manifestation of these impacts on the absolute amount of forest cover remains to be documented, but there is little doubt that militarization is changing the configuration of the forest.

Second, the government response to local needs in areas both influenced and not influenced by the Zapatistas has been quite dramatic. Roads that have been in planning for over a decade instantly became top priority, largely for security reasons. These roads, discussed in Chapter 4, have opened up remote areas to increased traffic and economic activity. Electricity, solar panels, and other amenities that have eluded the region for decades have suddenly been forthcoming. For example, by the end of 1997, 201 kilometers of transmission lines will supply electricity to more than 60,000 residents of Las Cañadas subregion (*El Nacional*, 8 February 1997). The repeal of the veda forestal was considered by many to be a concessionary act towards the residents of the Marqués de Comillas subregion. In short, some parts of the Selva Lacandona have clearly benefitted from the uprising. As a result, there has been greater activity, and in many cases an increase in the amount of deforestation.

Finally, and perhaps most important, land takeovers and invasions have increased in the Selva Lacandona region, particularly on lands belonging to the Comunidad Lacandona and located within the protected area of the Montes Azules Biosphere Reserve (Nations 1994). In the aftermath of the Zapatista Uprising, over 100,000 hectares of tropical forests have been lost, according to Gabriel Quadri de la Torre, president of the Instituto Nacional de Ecología (Cruz 1997). Most of the invaders are not Zapatistas, but campesinos taking advantage of the region's tense political situation. According to a Conservation International report:

> The Tzeltal Maya invaders appear to be taking advantage of political unrest in the region instigated by the armed rebellion of the Zapatista National Liberation Army that began in early 1994, destroying more than 200 hectares of Lacandon Maya forest in a northern section of the Selva Lacandona. Over the past year, the Tzeltal have taken over another 2,000 hectares of Lacandon forest. Their extreme actions reflect the intense

economic and land-use pressures that are crippling the region (Conservation International Web Page).

Once settled, it is extremely difficult for the government to remove squatters from invaded land. Private lands have also been taken over in the aftermath of the uprising, including forested lands that had been earmarked for conservation by their legitimate owners. These land invasions have created new challenges for conservation efforts in the Selva Lacandona. While the cultivation of several dozen hectares of forest may not seem dramatic within the larger region, one settlement often leads to another. With each land invasion, the countervailing pressures of conservation are weakened and the future of the Selva Lacandona becomes more precarious.

8

Conservation Strategies

The Other Part of the Picture

The previous chapters have focused entirely on the driving forces of deforestation in the Selva Lacandona. These forces help to explain why certain patterns of deforestation emerged during certain periods. However, an emphasis on the driving forces does not provide a complete picture of the situation in the Selva Lacandona. Like a photographic negative, deforestation highlights only a part of the picture. The other part, considered figuratively to reflect the "positive," consists of the area that is still covered by tropical forest ecosystems.

The patterns of remaining tropical forests in the Selva Lacandona region are dictated in part by the region's geography. Many mountain crests and steep slopes remain heavily forested, while riparian forests, valleys, and the surrounding slopes are increasingly cleared. In contrast, the large, contiguous tracts of tropical forest that form the core of the Selva Lacandona can be attributed not to geography, but to conservation efforts. Working with the Comunidad Lacandona, which holds title to much of this core area, scientists and environmentalists have succeeded in countering some of the pressures that have degraded or destroyed many other parts of the forest.

The countervailing pressures for conservation include the incorporation of an increasing amount of land into protected areas, restrictions on land use in these protected areas, and the logging ban that was declared in 1989 and lifted in 1994. To supplement these measures, a number of projects and programs have been initiated in the Selva Lacandona over the last two decades. Governmental institutions, private and public research organizations, local, national, and international conservation groups, and Mexican and foreign universities have initiated countless studies and projects concerning rural development, colonization, biological diversity, and conservation (Ecosur 1997). In fact, since the 1970s, over 50 large-scale projects have been initiated in the Selva Lacandona by many of the leading institutions of Mexico (Vásquez Sánchez 1992).

Preserving Mexico's largest remaining tropical forest has involved a continuous struggle to influence government policies, capture financial resources to support conservation, and convince local, national, and international constituencies of the importance and urgency of the effort. Over two decades of work has been marked by both successes and disappointments. The most notable successes involve the declaration of a series of protected areas in the Selva Lacandona. Presently, an estimated 412,236 hectares of land hold some type of protected status in the Selva Lacandona (Table 8.1). In addition, Mexico has been able to secure significant funds for conservation through the World Bank's Global Environmental Facility (GEF).

In spite of these successes, land invasions taking place within protected areas have frustrated efforts to guarantee the future of the Selva Lacandona. Since 1994, an estimated 100,000 hectares of tropical forests had been lost in Chiapas, including parts of the Montes Azules Biosphere Reserve (Cruz 1997). Furthermore, deforestation in the communities surrounding protected areas has put increased pressure on the reserves. As a result, it is becoming clear to many within the conservation community that management plans for protected areas in the Selva Lacandona must address the social struggles of the region's residents if they are to succeed. Consequently, a gradual reorientation of conservation strategies is underway, with the outcome as yet to be determined.

TABLE 8.1 Protected Areas in the Selva Lacandona region of Chiapas, Mexico

Name	Date Decreed	Size (ha)	Type
Montes Azules	1/21/78	331,200	Biosphere Reserve
Bonampak	8/21/92	4,357	Natural Monument
Chan-Kin	8/21/92	12,184	Protected Area of Flora & Fauna
Lacantún	8/21/92	61,874	Biosphere Reserve
Yaxchilán	8/24/92	2,621	Natural Monument
TOTAL		412,236	

Source: Ordoñez Díaz and Flores Villela, 1995.

Mexico's *Patrimonio Natural*

The Selva Lacandona is an indisputable biological treasure trove, and it forms an important part of Mexico's *patrimonio natural*, or natural heritage. The area within the Montes Azules Biosphere Reserve alone is home to a significant proportion of Mexico's biodiversity (Table 8.2). A number of species inventories undertaken by Mexican and international biologists attest to the rich species diversity of the Selva Lacandona (Vásquez Sánchez and Ramos Olmos 1992). It has been estimated that at least 4,000 plant species exist in the forest, including some which are endemic to the area. These include the epiphyte *Yucca lacandonica* (Gómez-Pompa and Valdes 1962) and a separate vascular plant family known as Lacandoniaceae, represented by a single species of *Lacandonia schismatica* (Martínez and Ramos 1989). The Selva Lacandona is home to at least 306 bird species (Iñigo-Elías 1991; González-Garcia 1992), and is also an important host for dozens of other migratory birds from North America (Greenberg 1990). In addition, the region contains one of the most diverse mammalian communities in Mexico. At least 82 species have been recorded, and another 42 are potentially present (March and Aranda 1992). Bats alone comprise a large portion of this diversity — at least 46 species have been recorded so far, distinguishing the region as having among the highest bat diversity in the world (Medellín et al. 1992). Twenty-three amphibian and 54 reptile species have also been identified (Lazcano-Barrero et al. 1992), along with at least 59 species

TABLE 8.2 Biological diversity in the Montes Azules Biosphere Reserve, Chiapas.

	Mexico	*Montes Azules*	*Representativeness (%)*
Types of Vegetation	47	13*	28
Flora	30,000	3,000*	10
Orchids	1,200	320	27
Mammals	439	106	24
Birds	961	306	32
Reptiles	717	84*	12
Amphibians	284	25*	9
Butterflies	1,800	800	44

* Estimates based on preliminary studies
Source: Agrupación Sierra Madre, no date.

of fish (March et al. 1997). At least 1,135 species of insects have been registered, representing a mere 3 percent of the estimated insect diversity of the Selva Lacandona (Morón 1992). These inventories, which are partial at best, underscore the need for further research, as well as the conservation of the Selva Lacandona.

Concern over the disappearance of the Selva Lacandona dates at least as far back as the 1940s, when colonization was just getting underway.[1] Early proponents of forest conservation included Gertrude Duby Blom and Miguel Alvarez del Toro (Woodward and Woodward 1985; Alvarez del Toro 1985; Simonian 1995), as well as a handful of scientists who had conducted research in the region. The Lagos de Montebello, located in the pine-oak forests on the westernmost edge of the Selva Lacandona region, was the first area in Chiapas to acquire protected status in 1958. The government's willingness to protect this area was in part inspired by potential tourism revenues.

Despite localized concern, no concrete measures were taken to protect the Selva Lacandona until 1972, when the area surrounding Laguna Miramar was declared an area of environmental conservation at the urging of Alvarez del Toro (Vásquez Sánchez 1992). In the same year, an influential article on tropical rain forests, written by three Mexican scientists, appeared in the journal *Science* (Gómez-Pompa et al. 1972). This article stressed the importance of tropical forests for biodiversity, and alerted the international scientific community to the danger of mass extinctions resulting from deforestation.

Pressures to protect the Selva Lacandona grew during the 1970s, as logging by COFOLASA razed more and more of the forest and the region became increasingly populated. However, it was not until December, 1977, when the first significant conservation measure for the Selva Lacandona was announced. At that time, President Lopez Portillo announced the creation of the 331,200-hectare *Reserva de la Biósfera Montes Azules* (RIBMA). The announcement came in response to lobbying by a consortium of institutions, including the *Instituto de Ecología*, the *Instituto Nacional de Investigaciones sobre Recursos Bióticos* (INIREB), the *Centro de Ecodesarrollo* (CECODES), and the *Centro de Investigaciones Ecológicas del Sureste* (CIES) (Vásquez Sánchez 1992). The Montes Azules Biosphere Reserve was officially decreed in the government's official legislative publication in January, 1978, along with a vast, 26,123 km^2 zone of forest protection extending across the upper basin of the Río Usumacinta, including parts of the Highlands of Chiapas (Diario Oficial 1978).

Montes Azules forms part of UNESCO's Man and the Biosphere (MAB) Project 8, which seeks to conserve land while simultaneously considering the needs of the local population through the creation of biosphere reserves. Research, education, monitoring, and rural development are included in the MAB model of national parks (Halffter 1980b). Ideally, biosphere reserves are organized into three interrelated zones, referred to as the core area, the buffer zone, and the transition area. The core area should not be subjected to human activity, with the exception of research and monitoring, and in some cases traditional extractive uses by local communities. The buffer zone, which is contiguous to the core area, comprises a zone of managed use, where human activites do not interfere with conservation objectives. The transition area is the outermost zone, where local communities, conservation organizations, scientists, civic and cultural groups, and other stakeholders work together to manage and sustainably develop the resources for the benefit of the people who live there (MAB Web Page).

From its inception, Montes Azules failed to incorporate the MAB ideals and organizational structure into its design. This may not have been an oversight, given that the biosphere reserve concept was in its infancy, and still evolving in terms of both objectives and operations. In fact, at about the time that Montes Azules was decreed, some confusion had arisen regarding how biosphere reserves related to other conservation efforts. It was only after the International Union for the Conservation of Nature (IUCN) and UNESCO published a paper in 1979 clarifying the relationships between biosphere reserves and other types of protected areas that a unique identity for biosphere reserves was established (Gregg and McGean 1985).

From another perspective, it seems that the failure to establish buffer and transition zones around Montes Azules was simply a mistake. Reflecting on the unique role of biosphere reserves in Mexico, Gonzalo Halffter, one of the key advocates of the biosphere reserve concept in Mexico as well as an early proponent of Montes Azules Biosphere Reserve, suggested that there was a discrepancy in the way that the reserve was conceived, and the way that it was carried out:

> Considerable pressure is being brought to bear for the biosphere reserves to be controlled in a similar way to the national parks, and as a result, certain legal documents run counter to the efforts of the scientists (particularly at the Institute for Ecology) and contain *mistakes* such as the designation of the whole area selected as a conservation zone, called an 'integral reserve' (e.g. the reserve of Montes Azules, Chiapas) (Halffter 1980a:272, italics added).

Reviewing one of the preliminary studies for Montes Azules undertaken by the Institute for Ecology under the auspices of the multi-institutional *Fideicomiso de La Selva Lacandona* (1977), it seems that the absence of zoning and the failure to incorporate local communities in the management activities of the reserve is more readily explained by an incipient and underdeveloped philosophy of biosphere reserves: Montes Azules was created as a reserve to protect Mexico's germplasm and biological patrimony, and to foster scientific and technological research on its potential uses for present and future generations.

The factors mentioned in the report as important for selection into the Man and the Biosphere program included both primary criteria, such as representativeness, diversity, natural state and efficiency as a unit of conservation, and secondary criteria, such as information about the zone, endangered species, historical importance, and so on (Fideicomiso de la Selva Lacandona 1977). Buffer zones and transition zones were *not* an integral part of the plans for Montes Azules.

The Fideicomiso's report acknowledged that the region was *not* untouched by humans. Maya ruins attested to its historical use, and more recent impacts were recognized, including the logging of mahogany and cedar trees, sparse settlements of Lacandón Indians, and several more recent settlements. However, the report stressed that human modifications were restricted in the reserve area, and that disturbed areas bordering the core of the reserve are capable of regenerating, or being put under "rational control" (Fideicomiso de la Selva Lacandona 1977:124).

Three communities were recognized to inhabit areas directly outside of the reserve limits: Benito Juarez, el Granizo, and Velasco Suárez. The area within the reserve was considered practically uninhabited, with the exception of a few families north of the Cañon del Colorado (located to the south of Laguna Miramar). Other settlements could be found to the northwest of the Sierra del Caribe, next to the Lagunas Lacanjá and Suspiro, along the Río Tzendales, and at the mouth of the Río San Pedro. Following the philosophy of the reserve, these inhabitants were not to be excluded, but rather incorporated into the plans. It was anticipated that the reserve could foster the development of rational uses of flora and fauna to successfully replace the traditional agriculture and ranching systems that were degrading the forest.

The published decree of January 12, 1978 stipulated that outside of the declared reserve, the use or exploitation of the forest and the fauna and economic development would take place in a manner that ensured the best and most sustainable use of the resources (Diario Oficial 1978). To comply with this, the *Secretaría de Agricultura y Recursos Hidráulicos* (SARH) was to formulate projects and establish norms for use of resources. The decree also stated that within the reserve, areas would be defined where the only permitted activities would be tourism and scientific or technological research. Agriculture and ranching were only allowed on already-deforested lands, or on *acahuales* (secondary growth forests) of less than 20 years old, subject to SARH's approval (Diario Oficial 1978).

Although Montes Azules was hailed as an example of the Mexican government´s "progressive and dynamic attitude" in the work of conservation and protection (Anonymous, 1980:32), it was not a model biosphere reserve in the sense of the MAB program. The policies followed in other Mexican biosphere reserves, such as Mapimí and La Michilía in Durango, were simply not followed in Montes Azules (Halffter 1984). According to Vásquez Sánchez (1992:22), the decree that established Montes Azules was set forth without prior technical studies and without consulting the local population. The decree merely defined the borders of the reserve, and set some normative guidelines for environmental protection. In Halffter's opinion, the problem was the lack of a scientific institution to promote conservation and establish a presence in the area (Halffter 1984).

As for the vast zone of forest protection declared for the Selva Lacandona in the 1978 presidential decree, it appears to have been quickly forgotten. In the public interest, the decree held that every type of resource use or exploitation was to be governed according to "properly based and applied scientific and technical norms" (Diario Oficial 1978:7, own translation). The decree did not mention that other interests were simultaneously driving thousands of colonizers into the very same forest, or that PEMEX had recently initiated oil exploration in the same area. Although all of the Selva Lacandona was legally "protected" in 1978, it was widely understood that the core of the protection was concentrated in the Montes Azules Biosphere Reserve.

For almost fifteen years, no additional areas came under protection in the Selva Lacandona. From 1978 to 1992, thousands of colonizers moved into the region to establish communities and small farms. Roads constructed by PEMEX, COFOLASA, and the Mexican military facilitated access to once-remote parts of the forest. The driving forces of colonization, discussed in the previous chapters, ensured a steady supply of migrants. It was not until the end of the 1980s that the future of the Selva Lacandona surfaced as a serious concern, both nationally and internationally.

In response to growing public attention and the efforts of the conservation and scientific community, several additional areas within the Selva Lacandona were legally protected in 1992. Some of these included sites of archaeological significance, such as the ruins at

Bonampak and Yaxchilán (Diario Oficial 1992; Diario Oficial 1992). A biosphere reserve consisting of 61,874 hectares of primarily tropical forest was also declared (Diario Oficial 1992). This reserve, known as Lacantún, is contiguous to the Montes Azules Biosphere Reserve, and also located on land belonging to the Comunidad Lacandona. Finally, a 12,184-hectare protected area called 'Chan-Kin' was declared between Frontera Corozal and Benemérito (Diario Oficial 1992). This area also belongs to the Comunidad Lacandona, and is thought to be devoid of human settlements. From a biodiversity perspective, these areas contribute to a forest corridor connecting the Selva Lacandona to the Petén region of Guatemala. Such biological corridors are considered essential to the conservation of many species.

Mexico's environmental ministry, SEMARNAP, has identified two other candidates for protected area status within the Selva Lacandona; Najaha and Metzabok (SEMARNAP 1996). Each of these would consist of 4,000 hectares of forests and lakes. Part of the rationale behind these protected areas would be to help create or reinforce a local consensus in favor of conservation (SEMARNAP 1996). However, it is unlikely that these lands will become integrated into Mexico's network of protected areas, given that the National Indigenous Congress recently adopted a resolution suspending all projects to declare natural protected areas within indigenous territories, including in the Selva Lacandona (Carlsen 1996).

The locations of existing protected areas are shown in Figure 8.1. From this map, it becomes clear that conservation efforts have been concentrated in only one part of the Selva Lacandona. The forests to the north, to the west, and increasingly to the south of this conservation zone have been tacitly surrendered to colonization, agricultural expansion, cattle ranching, oil exploration, and other activities. Any blocks of forests existing outside of protected areas are subject to the discretion of local communities and individual landowners. At the same time, the forests under protection are vulnerable to spontaneous colonization and agricultural expansion. Whether these forests are managed or cleared lies at the crux of the debate regarding the future of the Selva Lacandona.

Managing the Forest

The management of forests in the Selva Lacandona has been of vital interest to the conservation community, often resulting in a frustrating struggle to influence state and national policies. In the wake of the 1978

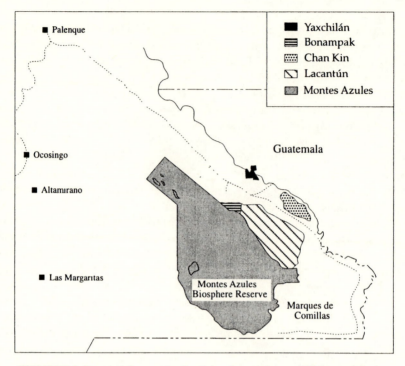

FIGURE 8.1 Protected areas in the Selva Lacandona of Chiapas, Mexico (*Source*: March, 1994).

presidential decree declaring the Montes Azules Biosphere Reserve, scientists and conservationists had little means of establishing a presence in and around the newly-protected area. Despite the creation of a directive and technical council, there was literally no budget for managing the reserve (Vásquez Sánchez 1992).

Institutional conflicts marked the early history of the Montes Azules Biosphere Reserve. Official management of the reserve was initially assigned to SARH's *Subsecretaría Forestal y de la Fauna* (Halffter 1984). With the creation of SEDUE (*Secretaría de Desarrollo Urbano y Ecología*) in 1982, control of the reserve shifted domain. This transfer initiated a conflict between INIREB and the environmental ministry, SEDUE, over control of the biosphere reserve. In 1984, INIREB succeeded in gaining administrative control over Montes Azules, and was assigned responsibility for directing research on the Selva Lacandona. It was also responsible for developing a program for Montes Azules that concentrated not only on the physical and natural environment, but also on social issues related to the use of the area's resources (Vásquez Sánchez 1992).

SEDUE had initiated the construction of seven stations for vigilance and control along the perimeter of the Montes Azules Biosphere Reserve in 1983, including a biological research station at the southern edge of the reserve. A number of buildings were constructed by 1984, but the interest and financial resources disappeared before they were completed. The sites were abandoned in 1985, and tropical vegetation took over the buildings. In 1987, the station was revived through the initiatives of two Mexican scientists, Rodrigo Medellín and Eduardo Iñigo-Elías. Their efforts led to a collaboration among the *Centro de Ecología* at the *Universidad Nacional Autonomo de México* (UNAM), the University of Florida, and Conservation International (Medellín 1991). Over the past seven years, the Chajul Tropical Biology Station has been successfully transformed into a research station for the *Instituto de Ecología* at UNAM (formerly the *Centro de Ecología*), with infrastructure and support supplied by Conservation International. For much of this period, the research station was the only visible sign of activity carried out in the Montes Azules Biosphere Reserve.

A *Comisión Intersecretarial para la Protección de la Selva Lacandona* was established in 1985, and included representatives from a variety of government ministries. Technical advisors from conservation groups and universities were also included on the commission. One of the goals was to work with ejidatarios to regulate the exploitation of timber resources. A number of ejidos signed agreements declaring part of their land as permanent forests that could be managed for timber extraction. However, like many other commissions, its accomplishments fell short of its original objectives (Vásquez Sánchez 1992).

In 1988, INIREB was closed down, and most research activities in Montes Azules Biosphere Reserve came to a halt. In late 1988, several biologists from INIREB formed a non-governmental organization called Ecosfera to continue conservation research in the Mor tes Azules Biosphere Reserve. Although short-lived, Ecosfera p oduced a significant amount of research on the Montes Azules Biosphere Reserve (see Vásquez Sánchez and Ramos Olmos 1992).

A series of other projects and management programs were initated in subsequent years, but rarely were the efforts coordinated among the numerous institutions working in the region (Vásquez Sánchez 1992). The projects resulted in a series of reports, essays, declarations, and high-level visits to the region. However, few of the projects succeeded in establishing a management strategy for the remaining tropical forests, integrating local people into conservation plans, or slowing deforestation. Heavy competition for scarce resources contributed to the demise of many ambitious and well-intentioned programs.

By 1990, conservation efforts had made only a small mark on the region. At the same time, the Selva Lacandona was rapidly changing, with new threats to the biosphere reserve emerging every year. With the help of environmental groups and the media, the Selva Lacandona began to take on growing importance. As a result, an international crusade got underway to save Mexico's last large rain forest.

By the late 1980s, tropical forests had come to the forefront of international environmental discussions, not only because of their rich biological diversity, but also because deforestation was seen as an important contributor of carbon dioxide to the atmosphere, which could impact the global climate system (Detwiler and Hall 1988). With a substantial base of support in their home countries, international NGOs became increasingly active in tropical forest conservation.

In the Selva Lacandona region, Conservation International (CI) of Washington, DC, has become the most visible international environmental organization involved in protecting the tropical forest. Competing organizations such as the World Wildlife Fund, which initially supported research in the Selva Lacandona, informally ceded control of conservation in the Selva Lacandona to CI. The driving mission behind Conservation International is "to conserve the Earth's diversity, and to demonstrate that human societies are able to live harmoniously with nature" (Conservation International Web Page). Since becoming involved in the region in 1989, CI and its Mexican affiliate, CI - Mexico, have attracted significant financial resources through two large debt-for-nature swaps as well as a substantial contribution from the Pulsar Corporation. It has used these resources to implement a series of "conservation solutions."

The first debt-for-nature swap with Mexico was arranged by CI in 1991. The organization purchased four million dollars of Mexico's debt at a 35 percent discount, then converted the amount to pesos to donate to conservation programs in Mexico. Approximately 2.6 million U.S. dollars were directed toward conservation efforts in the Selva Lacandona. These efforts included the following: ecosystem conservation data centers in the Centro de Ecología at UNAM and at institutions in Chiapas; the operation of the Chajul Tropical Research Station within the Montes Azules Biosphere Reserve; communication and education campaigns at the national and grassroots levels carried out by three private Mexican organizations, as well as the government environmental ministry (SEDUE, and later SEDESO) and the *Instituto de Historia Natural* in Chiapas (*Conservation International News Release*, 19 Feb. 1991).

In a second debt-for-nature swap negotiated in 1995, CI used $246,000 from the U.S. Agency for International Development to purchase $336,500 worth of Mexican debt held by a French bank. This amount was converted to pesos by the Mexican government, and the money was committed exclusively to conservation projects. As a result of the swap, CI generated $90,000 for its conservation work in the Selva Lacandona (Conservation International Web Page).

In 1996, a donation of ten million dollars was made to CI by Pulsar Internacional, S.A. de C.V. of Monterrey, Mexico to support the preservation of forests in the Selva Lacandona region. This donation was earmarked to "fund CI's programs that teach Selva Lacandona farmers how to support themselves through means that are conservation friendly. Farmers will learn to work the land in a buffer zone surrounding the forest with a particular focus on cultivating bamboo, or 'hule,' African palms and ornamental plants" (Conservation International Web Page).

In the years following the debt-for-nature swaps, CI's presence has increased substantially. It currently oversees a number of projects, all of which are considered to address deforestation. For example, CI maintains a constant presence at the Chajul and Ocotal field research stations, thereby discouraging environmental degradation of the surrounding area. It is also developing an ecotourism strategy for the area surrounding Laguna Miramar, running a women's embroidery cooperative at Laguna Ocotal, and providing technical support for Lacandonia, A.C., a private association formed by the Comunidad Lacandona in support of the sustainable development of the Selva Lacandona. Many of these initiatives involve some form of community development.

Over the past few years, the Mexican government has reviewed and revised its program on protected areas, culminating in the Program for Protected Natural Areas of Mexico, 1995-2000 (SEMARNAP 1996). The objectives of the program include expanding the coverage of protected areas and consolidating and promoting the adequate management of the system through effective mechanisms and efficient institutions (SEMARNAP 1996). The program involves a series of strategies, including the development of local institutions, the pursuit of opportunities for sustainable regional development, and budgetary increases through international financing, ecotourism, private investments, and other means. The new system of administration is to be carried out as pilot projects within 25 protected areas, including the Montes Azules Biosphere Reserve in Chiapas (SEMARNAP 1996).

In August 1997, a technical advisory council was established for the Montes Azules Biosphere Reserve. This *Consejo Técnico Asesor* (CTA) will seek to facilitate the participation of diverse groups in the management and conservation of Montes Azules (March 1997). The CTA is made up of representatives from communities, ejidos and social organizations, academic institutions, government, and NGOs. Anticipated to meet two times a year, the CTA will consider proposals for conservation, with an emphasis on active participation from different sectors of society and providing a forum for constructive criticism and debate.

The emerging emphasis on the diversified management of protected areas in Mexico can be considered another successful outcome of conservation efforts in Mexico. However, management strategies restricted to protected areas are unlikely to guarantee the future of the Selva Lacandona. Although such strategies may reinforce the countervailing pressures of conservation, the protected areas remain vulnerable to the driving forces of deforestation. Without addressing the complex web of social, economic, and political forces that have led to deforestation, environmental and social struggles will continue along distinct paths in the Selva Lacandona region.

Conservation and Development

The environmental community has increasingly recognized the importance of including local people in conservation plans. The partnership between conservation and development dates back to the late 1980s, when the Bruntland Commission report on "Our Common Future" stressed that environment and development are not separate challenges, but instead are inexorably linked (World Commission on Environment and Development 1987). Over the past decade, there has been a great deal of rhetoric about sustainable development in the Selva Lacandona, matched by numerous programs and projects that have by most accounts failed to achieve either conservation or development. Most of these projects, in fact, did not include monitoring and evaluation components in their design, thus it has been difficult to assess their impacts.

Some of the most common efforts to promote sustainable development have involved projects aimed at introducing different agricultural crops for production and marketing. Over the years, the government has sponsored projects for the cultivation of cacao, rubber, chile, and palm oil, among others. However, the projects have almost always failed, for lack of resources and follow-up, or simply because a crop unsuitable for local conditions was selected. This has led to a bitterness among campesinos, who feel that the government has been experimenting "with our sweat and our money" (Rojas 1995:125, own translation).

Some of the most critical development problems in the area were identified at a 1986 conference held at the Centro de Investigaciones Ecológicas del Sureste (CIES, now *El Colegio de la Frontera Sur*, or Ecosur). These problems included land tenure conflicts, insecure landholdings resulting from the non-execution of presidential resolutions; lack of institutional coordination among representatives of the agrarian ministry and campesino organizations to support development; the use of exemption certificates for cattle ranchers; and a flow of immigrants resulting from hydroelectric and petroleum projects. A series of alternatives were proposed, ranging from the reorientation of agrarian policies to the identification of possible or eventual areas for colonization, not necessarily in the Selva Lacandona region (Rojas 1995).

Addressing land tenure issues in the Selva Lacandona region would involve radical changes in agrarian structures that have persisted for generations. Not surprisingly, very few of the alternatives proposed at the 1986 CIES conference were pursued. Instead, development strategies continued to focus on poorly coordinated and heavily underfinanced projects aimed at generating income on already-established ejidos or communities.

From the campesinos´perspective, the government has failed to live up to its promises, over and over, regarding conservation and development. For example, in 1987 SEDUE promised to send food and resources to assist locals in the protection and management of the forest, but nothing ever came of it. This contributed to deep cynicism regarding conservation. In a 1988 article in *La Jornada*, Rojas (1995:127, own translation) captured the prevailing attitude among campesinos toward conservation:

> The bureaucrats of Sedue brought us a box of postcards from children from Europe asking President de la Madrid to protect the Selva Lacandona, but this can only be accomplished with development for the people who live here. In whatever ministry of government, in whatever program, they tell us that there is no money, that the country is in a crisis. Well then, if the international community wants to save the forest, why don´t they bring the money to invest in the Selva?

Recently, efforts to establish a post-facto buffer zone around the Montes Azules Biosphere Reserve have been initiated by a Mexico City-based environmental organization called Montes Azules, whose motto is "Conservation through Management." This organization, supported by Conservation International, views the current social realities in the communities bordering the reserve and the sustainability of their economic activities as particularly dismal for conservation:

> In the absence of viable economic alternatives in an area suffering nearly 100 percent unemployment, local populations are utilizing the forest (including the Montes Azules Biosphere Reserve) in an unsustainable manner for such subsistence purposes as slash and burn agriculture and timber (Montes Azules Web Page).

Seeking to enable local communities to "socially and economically develop in harmony with the rainforest," the organization's objective is to create environmentally sensitive jobs in buffer zone communities (Montes Azules Web Page). For example, Montes Azules oversees a butterfly ranching operation in the ejido Boca de Chajul, along with mushroom farming projects, the production of embroidery products, woodcrafts, essential oils, and ecotourism in communities surrounding the biosphere reserve. Although the products are marketed to a small but growing number of tourists in the region, Montes Azules also advertises its rainforest products through its homepage on the World Wide Web. An affiliated non-profit organization, *Espacios Naturales y Desarrollo Sustentable A.C.*, is supporting a scarlet macaw conservation project in the ejido Reforma Agraria (Montes Azules Web Page).

While the butterfly ranching project is touted as the only legal such project in Mexico, its impact on the Boca de Chajul community has yet to be evaluated. The project's impact on butterfly biodiversity has also not been assessed. As of late 1996, many of the collectors were small children who were paid several pesos for each butterfly that they collected. There was, however, a large educational component to the project, and one could argue that the visibility of the project and its educational impacts are of greater importance than its economic consequences. Nevertheless, in terms of sustainable development, it is unlikely that butterfly ranching will replace agriculture as the predominant economic activity in Boca de Chajul or other communities in and around the biosphere reserves.

Ecotourism is considered by many to offer the greatest potential for sustainable development in the Selva Lacandona region. Indeed, striking scenery and rich biodiversity mark the Selva Lacandona as a popular destination for backpack travelers and avid bird watchers. To date, however, few ecotourism projects have been developed in the Selva Lacandona. While a large number of tourists visit the Maya ruins of Yaxchilán and Bonampak, few spend more than one or two days in the area. Exceptions include cultural tourism to the Lacandón communities of Naha and Metzabok, and a project in the ejido Reforma Agraria in the Marqués de Comillas region. In addition, Conservation International is supporting the development of ecotourism around Laguna Miramar, and along the Río Lacantún, at the mouth of the Río Ixcán. An ecotourism project around Bonampak is aimed at benefitting the communities of Lacanjá Chansayab, Bethel and San Javier. A road to this area was scheduled for completion in 1997, after which time the number of tourists is expected to surge.

There is little doubt that ecotourism in the Selva Lacandona region will expand dramatically in the coming years. Although the development of an ecotourism industry offers many opportunities for blending conservation and development, it should be explicitly recognized that tourism in general is an unstable source of income, influenced by uncontrollable factors such as political stability, climate, and international exchange rate fluctuations. Furthermore, a large part of the revenues often disappears into the pockets of tour operators or external suppliers (Boo 1990). The successful development of ecotourism in the Selva Lacandona is considered particularly challenging, not only due to the remoteness of the region, but also because of security concerns.

Since 1995, SEMARNAP has initiated a series of regional programs for sustainable development in Mexico. Through the *Programa de Desarrollo Regional Sustentable* (PRODERS), the Selva Lacandona region has become the focus of strategies aimed at decentralizing administrative and policy decisions regarding environmental management and regional development (PRODERS Web Page). Although the proposed strategies are quite general, the process itself seeks to involve representatives from different actor groups and establish a regional consensus regarding alternative paths of development.

To date, most sustainable development projects in the Selva Lacandona region have focused exclusively on one or two resources at a time, rather than seeking to diversify activities and establish alternatives to conventional production processes. In concentrating efforts on one or two crops, such as rubber or chile, most projects have been highly risky at the outset. At the same time, there has been little emphasis on "low risk" projects that require small investments and little infrastructure.

While much has been made of the failure of sustainable development projects in the Selva Lacandona region, a number of projects have been successful in elevating the incomes of individuals or small groups within a community. For example, Arizpe (1996:15) describes how the introduction of chile as a commercial crop helped to make some farmers in the Selva Lacandona region extremely wealthy. These individuals, who were generally involved in commercial agriculture and cattle raising to begin with, were able to secure the support of campesino groups such as the *Uniones de Ejidos* and prosper from the sustainable development initiative. Arizpe notes that this economic differentiation exacerbated tensions in the region, as impoverished residents felt excluded from the new source of wealth.

"Sustainable development" is an appealing solution to contemporary challenges, but very few of these projects have proven to be successful in the Selva Lacandona on a scale that is meaningful to local economies. Although raising the income of a few individuals may render a project successful in the eyes of donors or project managers, the net impact on deforestation is likely to be quite small, unless the underlying forces are addressed at the same time.

Tensions over Tenure

The conservation activities carried out in the Selva Lacandona have not been without controversy. The borders of Montes Azules coincided with much of the 614,321 hectares of land given to the Lacandón Indians in 1972 (Diario Oficial 1972). In fact, 86 percent of the reserve (238,773 hectares) belongs to the Comunidad Lacandona. Moreover, a number of ejidos had already been established within the defined reserve, such that 47,427 hectares were legally considered ejidal lands prior to 1978 (March 1994). Additional ejidos were titled in subsequent years. As a result, by 1990 there were as many as 39 settlements within the reserve, with a total population exceeding 3,500 (March 1994). Some of the contradictions between conservation goals and land titles can be seen in Figure 8.2. A number of areas have been designated as part of the Comunidad Lacandona, as part of the Montes Azules Biosphere Reserve, and as ejidal lands. These overlapping claims have inevitably led to tensions and conflicts.

In 1985, Miguel Alvarez del Toro forewarned that unless the environmental ministry, SEDUE, took action, the *Secretaría de la Reforma Agraria* (Ministry of Agrarian Reform) would continue to distribute land within the Montes Azules Biosphere Reserve (Rojas 1995). Over the past twelve years, the conflicts have become more acute. Not only are there old claims remaining to be settled, but new claims are being made by spontaneous colonizers who have invaded the reserve.

FIGURE 8.2   Land use contradictions in the Selva Lacandona, 1990. The darkest areas are titled to both ejidos and the Comunidad Lacandona; the shaded areas that fall within the dashed line coincide with the protected lands of the Montes Azules Biosphere Reserve (Adapted from Lazcano-Barrera et al., 1992).

In 1988, over 3,000 campesinos from nine ejidos were in the process of being relocated by SEDUE and the agricultural ministry, SARH, in the name of preserving the Selva Lacandona (Rojas 1995). The resettlement involved a dispute over 6,800 hectares of land that legally belonged to the Comunidad Lacandona, but which in some cases had been settled *prior* to the 1974 resolution which established Lacandón ownership. These campesinos had resisted earlier government efforts to relocate them into concentrated settlements, claiming difficulties in accessing water and reaching their fields from the new settlements. The expelled campesinos were offered 10,000 hectares in the Marqués de Comillas subregion, in return for vacating the land under dispute. Critics of the resettlement program proposed that the agreement with the Comunidad Lacandona be revised, reducing their land area by more than 100,000 hectares (Rojas 1995).

The fact that the Montes Azules Biosphere Reserve is superimposed on the land of the Comunidad Lacandona has forced the environmental community to consider the interests and opinions of the 13,461 members of the Comunidad Lacandona. Made up of an alliance of three ethnic groups, the Comunidad Lacandona has emerged as the largest landholding group in the region. According to Nations (1994:33), the

Biosphere Reserve has strategic significance for the Comunidad
Lacandona: "Rather than viewing this overlap of protected area and
indigenous territory as a threat, the Tzeltal, Chol, and Lacandon Maya
of the Comunidad Lacandona see the Montes Azules Biosphere Reserve
as a buffer against outside threats to their land." At the same time, the
Comunidad Lacandona is unwilling to allow decisions and control of
their land to be usurped by environmentalists, and has increasingly
voiced its opinion in the management of the Selva Lacandona.

In 1992, scientists and conservationists proposed a 105,314-hectare
biological corridor that would link the tropical forests of the Selva
Lacandona to those of the Petén region of Guatemala (Gómez-Pompa
1992b). Since the proposed corridor was legally held by the Comunidad
Lacandona, the initiative was put forth to representatives of the
community for approval. In February 1992, representatives from the
three ethnic groups included in the Comunidad Lacandona met with
delegates from SEDESOL and SARH to discuss the proposal. The
Comunidad Lacandona came out *against* the establishment of an
official reserve by decree, instead favoring the pursuit of conservation
alongside carefully managed development. In 1993, the general
assembly of the Comunidad Lacandona established a communal reserve
known as "la Cojolita." This reserve of 13,165 hectares complements the
four national protected areas established in 1992, contributing to a *d e
facto* biological corridor between Mexico and Guatemala.

Discord between the Comunidad Lacandona and the environmental
community became visible in 1996, when the Comunidad Lacandona
criticized the criteria used by the Instituto Nacional de Ecología (INE)
to initiate projects in the Selva Lacandona. More than a simple
disagreement, the dissent caused the World Bank to suspend the
disbursement of 20 million U.S. dollars to the Mexican government
(Balboa 1996). This money, destined to be distributed among 10
ecological zones in Mexico, including two in Chiapas, was donated by
the Bank's Global Environmental Facility (GEF).

The Comunidad Lacandona opposed the distribution of GEF funds
through the government and private sector. Instead, they argued that
this money, earmarked for ecological reserves, should go directly to
local communities so that vast amounts do not get consumed in
administrative costs. The Comunidad Lacandona accused INE of
"bureaucratizing" the funds. The lack of a long-term vision for the
region, compounded by an absence of monitoring and evaluation
programs and the political nature of disbursements have, in the words
of the Comunidad Lacandona, turned the Selva Lacandona into a
"cemetery of failed projects" (Balboa 1996).

The use of public investments in rural or indigenous zones of Chiapas was seen by the Comunidad as a way of rewarding or punishing indigenous communities, their organizations or leaders, municipal presidents, political delegates, and religious leaders who had influence in decisions for the rest of the community (Balboa 1996). The government and conservation community responded to the Comunidad's concerns, no doubt to ensure that the GEF money was forthcoming.

In February 1997, officials from INE convened a meeting with the Comunidad Lacandona, including the communities of Nueva Palestina and Frontera Corozal, to discuss the management plan for the Montes Azules Biosphere Reserve and to organize educational workshops to integrate local communities in the management of the biosphere reserve. The Comunidad Lacandona requested that representatives from the World Bank be present at the workshops and meetings, in order to ensure compliance with the Bank's Sustainable Development Program and its Plan for Indigenous Development. Other national and international NGOs, such as World Wildlife Fund and Pronatura, were also invited to participate as observers. In follow-up meetings organized by the Comunidad Lacandona, each community designated representatives to participate in the workshops, and it was agreed that the *Comisión Intersecretarial para la Comunidad Lacandona* would meet every two months with members of the Comunidad to review the different projects taking place within their territory. Whether the concerns of the Comunidad Lacandona are taken in earnest by the conservation community remains to be seen.

The Logging Divide

The conservation community has been actively lobbying the Mexican government to take action to halt the imminent destruction of the Selva Lacandona. Influential environmental organizations, such as the *Grupo de los Cien*, a coalition of artists, writers, and celebrities, have successfully drawn attention to the plight of the Selva Lacandona, raising the public's awareness and placing it high on the government's priority list for environmental actions.

One of the conservation community's greatest political successes was the *veda forestal*, discussed in Chapter 4. This ban on logging appeared to mark the government's commitment to halting the destruction of the Selva Lacandona, as well as other forests in Chiapas. In the words of Patrocinio González Garrido, the governor who enacted the measures:

We have begun in Chiapas the end of the mirage of the abundance and exuberance of our tropical rainforest; concluded also is the Lacandon rainforest as an alternative to agrarian and social problems; it is likewise ended the renting of concessions in the rainforest, that never favored the inhabitants. . . . also terminated are the changes in land use, pretext for massive fellings that left us thousands of unproductive hectares. . . (quoted in Bray 1997:7).

The veda forestal was considered by many as the first concrete action to indicate that the government was serious about forest conservation. Nevertheless, "the lack of alternatives led campesinos to continue to cut timber for domestic use and for the illegal timber trade. Instead of resolving the issue, the ban only created political problems as campesinos and police faced off in violent battles over rolls of timber and bundles of firewood" (Harvey 1997:15).

It is no surprise that the decision to reverse the veda forestal in 1994 angered environmental groups, who claimed that 80 percent of the wood being removed was fresh wood, cut down specifically to sell in the market. SARH was also charged by environmentalists as having distributed 200 logging vehicles to those with permits, "creating an army of massive destruction" (*The News*, 20 Oct. 1994). As discussed in Chapter 4, the logging that took place after the ban was lifted did indeed involve the cutting of fresh wood, and logging exceeded the amount allowed by permits by at least five-fold.

The Comunidad Lacandona also came out against the reversal of the veda forestal, citing the following reasons: 1) the present forestry and environmental legislation contain serious deficiencies in the authorization of permits, and control over the utilization and the transport of the product, resulting in anarchy and chaos; 2) there is no forestry inventory in Chiapas; 3) government agencies completely lack the human and material resources for an efficient coverage and vigilance of forest exploitation; 4) there are no effective mechanisms in place to prevent the pillage of forest reserves; and 5) lifing the ban would greatly endanger the existence of protected areas within the Comunidad Lacandona (Villagran 1995).

The Comunidad Lacandona maintains that it is not against forestry *per se*, but stresses that it should be carried out in the context of community forestry compatible with sustainable development goals. The components of successful community forestry include;

. . . promoting the organizing of owners of forest resources into peasant enterprises that with government support achieve the integrative means for all the activities from the cutting, to the planting of trees and the means of transforming the industrialization of lumber, avoiding the concentration of wealth of these activities in a few hands and increasing the social class chasms and the resultant increase in poverty (Villagran 1995).

Community forestry has never been established in the Selva Lacandona region, and the repeal of the *veda forestal* has thus aroused fears that haphazard timber extraction would continue to benefit a few at the expense of a rapidly diminishing tropical forest. The controversies surrounding the logging ban serve as an important example of the distinction between environmental and social struggles in Chiapas, and the divisiveness of environmental issues in a rapidly changing frontier region.

Analyzing the controversy over logging in the Marqués de Comillas region, Harvey (1997) argued that the lack of alternative sources of income when the logging ban took hold created new pressures on the forest, making illegal logging the most economically ·viable option for local people to pursue. Focusing on the "web of power relations that operate through local agencies of environmental protection and 'development' projects," Harvey (1997:4) identified two opposing factions in the conflicts over logging. Not surprisingly, these factions are reflections of opposing environmental and social struggles in the Selva Lacandona:

> The main division within the state occurred between the Subprocuraduria de Recursos Naturales of PROFEPA [Procaduría Federal de Protección Ambiental] and the state delegation of SEMARNAP. We can think of these two agencies as adopting the position of the "good cop" and the "bad cop." The strategy of the good cop is to integrate ejidos into a legally regulated timber market. The bad cop prefers to discipline and punish. It is concerned about the lack of authority of the state on this frontier and its priority is to teach a lesson to those "unruly clients" who shared in illegal logging operations (Harvey 1997:20).

The "good cop - bad cop" analogy captures the dichotomy underlying social and environmental struggles in Chiapas. It reflects a growing sense of polarization, whereby one is either for the environment or for development. Finally, it reinforces the image of environmentalists as exclusionary, concerned more about the protection of flora and fauna than with human well-being. The fact that community forestry has not been developed in the Selva Lacandona represents a missed chance for generating income opportunities for local people directly from the forest (Bray 1997). The recently initiated *Plan Piloto Forestal*, modeled after a successful project in Quintana Roo, is aimed at reversing these trends. The success of such a project is critical, as campesinos have been willing and ready to participate in such a program for over a decade. If the permits do not come through, or if resources for the program are suspended, there is no doubt that the disappointment and disillusionment will reinforce the rift between environmental and social struggles.

The Future of Conservation

The Selva Lacandona is considered by the Mexican government to be among the highest conservation priorities in the country. This emphasis has led to the creation of new commissions, panels, and programs described above. The attention to the region is likely to grow in the future, as more of the remaining forests are transformed to agricultural lands or pastures, increasing the pressures on Mexico's biodiversity.

To date, the conservation community has achieved some notable successes in terms of the number and size of protected areas within the Selva Lacandona. It has also proven to be quite adept at securing international funding for conservation projects. Overall, the global importance attributed to the struggle to conserve the Selva Lacandona has fortified its role as a legitimate force to contend with in the struggle over the future of the Selva Lacandona.

Indeed, in the two decades since the Montes Azules Biosphere Reserve was decreed, the conservation community has become much more professional and politically savvy. Through experience, it has learned how to successfully negotiate with the government, international aid donors, and members of the Comunidad Lacandona. At the same time, the goals of environmentalists have become more realistic:

> It would be fruitless to argue for the preservation of all the forests of eastern Chiapas. Most of those forests are already gone. But it makes sense, for the benefit of the indigenous inhabitants of the Selva Lacandona and for the people of Mexico, to keep alive what little forest remains. . . The challenge remains to transform the rest of Chiapas into an ecologically sustainable mosaic of food production, small-scale cattle production, and extractive forest reserves (Nations 1994:33).

The tools and resources available today are far superior to those that existed two decades ago. Conservation organizations and scientists now have access to substantial amounts of information about the Selva Lacandona, and can monitor changes with relative ease and accuracy. Satellite images and geographic information systems with large social and physical databases have become the standard for assessing the state of the tropical forest. For example, Selvanet, an information system for the sustainable development of the Selva Lacandona region developed at Ecosur, consists of 61 digital coverages, along with a bibliographic database, lists of flora and fauna, and a climate database (Ecosur 1997). This system will support the government's sustainable development program, PRODERS, and eventually be made available to the public over the Internet.

With over 400,000 hectares in biosphere reserves and natural parks, more than 20 percent of the Selva Lacandona is currently under protection. Yet deforestation continues in the Selva Lacandona, and it is occurring on both sides of the borders of protected areas. Within the Montes Azules Biosphere Reserve, an estimated 10 percent (33,000 hectares) had been cleared by 1990 (Lazcano-Barrero et al. 1992). Invasions of protected areas have increased dramatically since the Zapatista Uprising, adding to the deforestation within the reserves. Peasants searching for land to cultivate are no longer intimidated by the Biosphere Reserve, and are moving in rapidly to establish settlements. The army has also invaded the area, in an effort to increase its presence for national security reasons. Meanwhile, the Selva Lacandona region has also become a center for smuggling, with drugs entering Mexico from Central and South America, and also cultivated within the region.

In the Selva Lacandona region, the approach of setting aside large tracts of land is being increasingly challenged as social struggles mature and the region becomes increasingly populated and integrated into the Mexican and global economies. Land invasions and encroachments into protected areas diminish the countervailing pressures of conservation. The conservation community has recognized the importance of involving local populations in its effort to protect the region's remaining forests. Nevertheless, those involved in social struggles are skeptical to the objectives of environmental struggles in the Selva Lacandona region. This skepticism can be considered the legacy of failed projects and unfulfilled promises. The major successes of environmental struggles have enclosed land from settlement and eliminated the possibility of community benefits from forest resources. Community-sponsored conservation plans have rarely been supported, despite initial enthusiasm from external institutions. Finally, there is a widespread feeling that conservation serves the needs of the international community more than the local people.

The future of conservation in the Selva Lacandona will depend on erasing local skepticism and involving residents directly in conservation efforts. However, to truly conserve the tropical forests of the Selva Lacandona, the underlying forces of deforestation will have to be simultaneously addressed.

9

Sacrificing the Forest

Driving Forces, Countervailing Pressures

The Selva Lacandona is not a homogeneous entity, and the driving forces of deforestation and countervailing pressures for conservation have contributed differentially to the present-day configuration of tropical forests in the region. Examining the political ecology of deforestation from a geographical perspective helps to explain how this configuration of forest cover came about, and why settlement patterns, ejidal structures, and protected areas have taken on their particular shapes.

The growing recognition that deforestation reflects the outcome of a complex array of factors can be considered a paradigm shift in understanding and explaining the causes of tropical deforestation. The direct causes of deforestation in the Selva Lacandona, including colonization, logging, petroleum exploration, and the expansion of both the agricultural frontier and cattle ranching, are clearly valid explanations for the continuing loss of forest. However, an analysis of the underlying causes indicates that a much deeper web of relationships is responsible for forest loss in this diverse region.

Through the analysis presented in the previous chapters, it becomes clear that the destruction of Mexico's largest remaining tropical rain forest reflects the way that production relations have played themselves out in the Selva Lacandona, with the actual circumstances varying both spatially and temporally. In fact, the Selva Lacandona has been sacrificed to the same economic, social, and political relations that have created an increasingly marginalized population in Chiapas. It should be no surprise that the land use decisions made by disparate actors in the Selva Lacandona follow the logic of these relations.

The driving forces of deforestation and the countervailing pressures of conservation have been distinguished here, not only because they differentially determine the configuration of tropical forests in the region, but also because they have very distinct implications for the future of the Selva Lacandona. The countervailing pressures exerted by conservation efforts are generally tied to activities taking place *within* the region. The growing recognition that environmental struggles must integrate communities in projects and support local conservation initiatives reflects an effort to build a shared vision of the future of the Selva Lacandona. The driving forces of deforestation, on the other hand, are largely tied to activities taking place *outside* of the region. These external forces, which are manifested through production activities as the direct causes of deforestation, are of key significance in securing the future of the Selva Lacandona, one that matches the vision developed through conservation efforts.

A number of the so-called solutions to deforestation have addressed the direct causes without taking into account the driving forces. The veda forestal, reduced credits for cattle ranching, environmental education programs, the voluntary withdrawal of PEMEX, and other measures have had little impact on deforestation rates in the Selva Lacandona. Forest loss continues, largely because some groups perceive no other options, while others perceive substantial profits. As an alternative to these "solution-oriented" strategies to halt deforestation, some authors have stressed that containing deforestation demands more comprehensive changes, including changes in social and economic relations.

According to Vandermeer and Perfecto (1995), addressing the question, *Why are there landless peasants?* is key to deterring the destruction of tropical forests. It also serves as a foundational question for the politics of social justice. Consequently, "the movement to save rain forests needs to be closely linked with, if not virtually the same form, as the movement for social justice" (Vandermeer and Perfecto 1995:159). The argument is specifically targeted to the Selva Lacandona region:

> If the people who live around the Lacandon Forest in Mexico, for example, have as their major goal the reformulation of the Mexican political system, the rain forest conservationist must join the political movement to change that system — something that many would see as distant from the original goal of preserving rain forests (Vandermeer and Perfecto 1995:167).

Faber (1993:234) also argues that "[a] new social order that raises the standard of living for all citizens is essential to comprehensively address problems of poverty, injustice, and environmental destruction." Weinberg (1991:163) goes even further to suggest that a "revolutionary environmentalism" is necessary to address environmental degradation in Central America:

> Revolutionary environmentalism would not merely take the mainstream movement's assumptions to a more radical level, but dispense with them completely in favour of a view that sees profound social and political transformation as both a principal instrument and an inevitable result of preserving wild areas and restoring the ecology.

Many analyses based on political ecology suggest that more equitable distributions of income, increased social justice, and greater democratic participation in government are among the necessary changes that can secure the future of tropical forests. Although such an analysis makes intuitive sense, it remains to be seen whether changes in social and economic relations will, in fact, work in favor of tropical forests. One possibility seldom considered is that the broad changes advocated by social activists or radical environmentalists will create new pressures for deforestation. While society at large may be better off as a result of the changes, tropical forests may still be lost. The outcome depends largely on the shape of the changes, and how they influence land, labor, and production relations. More important, it depends on whether conservation falls within the broader agenda for social change.

The Aftermath of the Zapatista Uprising

Calls for a new social and economic order within Mexico is precisely what the Zapatistas have demanded since the uprising began in 1994. The EZLN has openly challenged existing relations in Chiapas, in Mexico, and in the international system, questioning the basic validity of neoliberal models of economic growth. Since March 1994, the EZLN has been negotiating with the government for broad social, economic, and political changes, including autonomy for indigenous groups. The outcome of the negotiations between the Zapatistas and the government is presently unresolved. However, it is clear that the future of the Selva Lacandona is intimately tied to these negotiations. The future of the forest can only be anticipated by considering the demands of the Zapatistas, as well as the reactions of the government to these demands.

Although the Zapatista Uprising can be considered highly significant in the context of modern Mexican political history, the link between social, economic, and political changes and the protection of tropical rain forests is tenuous. Given that the underlying causes of deforestation are closely related to the causes of the rebellion, such changes can *potentially* benefit rain forest conservation. Nevertheless, a closer analysis of the situation suggests that a linkage between social and environmental struggles has not been forged in the Selva Lacandona.

Over the past four years, the Selva Lacandona has served as a stage for the peasant-based rebellion. The image of a tropical jungle has added to the mystique of the rebels, and provided an easily romanticized setting for both conflicts and negotiations. Yet, while the forest offers a dramatic backdrop for the Zapatistas, there has been virtually no mention of deforestation or ecological issues during the rebellion. Amidst the countless communiqués issued by the Zapatistas over the past several years (García de León et al. 1994; García de León and Monsivais 1995; Zapatista Communiqués Web Page), there has been a curious silence about the future of the tropical forest.

Perhaps this silence is not surprising, considering that deforestation is not a major concern among peasants struggling to make a living in the humid tropical environment. Townsend et al. (1995) interviewed pioneer women in the Selva Lacandona and heard stories of increasing poverty, lack of public services, and domestic conflict and violence. In contrast, they heard little about deforestation (Townsend 1996). In a similar study, Arizpe et al. (1996) interviewed residents in and around Palenque and within the Selva Lacandona region about their views on global change, including deforestation. Like Townsend et al. (1995), they found that deforestation is not perceived as a problem by many in the region, largely because "the issue has come from the outside and because no detailed studies or reliable data exist on its negative effects on agriculture, health, and other main concerns of local inhabitants" (Arizpe et al. 1996:94). Although the effects of deforestation were visible in certain parts of the region, no social awareness had been created locally.

Field interviews confirm that forest conservation is most often of secondary importance relative to survival, and more specifically, to economic betterment: "Protecting the rain forest, even by those willing to accept its value, is overridden by the indiscriminate search for profit, which is a main source of social and political prestige" (Arizpe et al. 1996:99). In the words of one Zapatista: "Ecologists? Who needs them? What we want here is land, work, and shelter" (Ross 1995:265).

Forest conservation has not been visible on the Zapatista's agenda, largely because it is not on the agenda. The Zapatistas struggle for democracy and justice has been underscored by a demand for land. In fact, the issue of land has been a source of conflict and violence in the region for over 20 years, and is considered to be one of the main causes of the Zapatista Uprising. Included with the demand for land was a list of additional inputs, including machinery, fertilizers, insecticides, credits, technical assistance, improved seeds, and cattle (*La Jornada*, 3 March 1994).

While some of these inputs are reflective of true agrarian reforms, others are precisely those that have been promoted as part of the Green Revolution in agriculture, and which have led to two-tiered agricultural development in Mexico. Satisfying these demands would involve an increase in government and private investment in the region, and the development of the region would be contingent upon improved infrastructure and services. As the history of the region has shown, an increase in such activities most likely will lead to a decrease in the amount of forest cover. Paz Salinas (1994:96) notes the irony of this situation, whereby the same development model that caused so much damage to Mexico's tropical forests continues to dominate under the questionable name of progress.

The issue of land is at the heart of current social struggles in the Selva Lacandona. As such, the current situation in Chiapas is disquieting in terms of the future of the Selva Lacandona. Approximately 70 percent of private property in Chiapas is currently protected from takeovers through exemption certificates and other legal mechanisms. This limits the possibility of using them for future agrarian reforms. Reyes Ramos (1992:119) points out that if agrarian reform is to continue, it will have to involve national lands that are becoming increasingly scarce. A lingering question is whether the government will maintain its commitment to preserving the tropical forests of the Selva Lacandona, or whether it will furtively use the remaining forests as an escape valve to satisfy a continuing demand for land.

Although it is difficult to speculate about the outcome of the Zapatista Uprising and its implications for the Selva Lacandona region, it is clear that unless the outcome significantly alters the driving forces of deforestation, then it is unlikely that changes will benefit the forest. The driving forces originate in social, economic and political processes operating at different scales. Given the complexity of the driving forces, and the difficulty in pointing to one single cause of deforestation, it is essential that social changes advocated by the Zapatistas explicitly take into consideration the future of the Selva

Lacandona. Until the concerns of conservationists are embedded in social struggles, it is likely that deforestation will continue in the name of social progress.

The Future of the Selva Lacandona

The Selva Lacandona has changed dramatically over the past three decades. However, these changes are being eclipsed by the rapid transformations that are taking place in the Selva Lacandona today. The region is being opened up to the outside world through new networks of roads, increased communications, and a flow of curious international tourists. Once a remote and difficult region to access, the Selva Lacandona is becoming increasingly integrated into a changing world.

Most analyses of deforestation end on a positive note, outlining a plan of action or recommendations that can mend the deficiencies in previous policies toward rain forest conservation. Such sanguine conclusions reinforce the idea that solutions to deforestation are readily available and straightforward, constrained only by political will, financial resources, lack of education, or the predominance of philosophies counter to conservation.

The analysis presented in this book paints a more somber picture of deforestation in the Selva Lacandona. From speculators in Mexico City who acquired title to large forested areas for colonization schemes at the turn of the 20th century, to campesinos in border communities who took advantage of cheap labor provided by Guatemalan refugees in the 1980s, there is no single culprit responsible for deforestation. Although the impacts of conservation efforts were recognized, their value as long-term solutions was questioned in light of the nature and extent of the driving forces of deforestation. Protected areas are becoming the focus of increasing conflicts, as more and more peasants invade the land to establish new settlements.

The distinction between environmental and social struggles in the region is clearly a major impediment to effectively addressing deforestation. Yet the common call for environmentalists to support social movements as a way of conserving tropical forests is questionable as an alternative solution. The analysis presented here suggests that it is equally important for social struggles to explicitly address environmental concerns. Unfortunately, the interests of national and international *ecologistas* and environmentalists are often viewed as inconsistent with the needs of the local people (Arizpe 1996). The struggle to preserve the biologically rich and visually spectacular forest for future generations remains quite distinct from the struggle to survive and prosper in a challenging physical environment. This distinction has created a polarized situation, whereby the two struggles are seen as competitive rather than complementary.[1]

If tropical forests are to survive into the next millenium, it is essential to identify a common ground between the two struggles. Whether the common ground lies in community forestry, ecotourism, agroforestry, or other income-generating options will probably depend on the communities and individuals involved. The resolution of land tenure conflicts may also be a critical element in dissolving the distinction between the two struggles. Just as the driving forces of deforestation affect the region differentially, the common ground is likely to vary dramatically across communities.

One point that has been stressed by the Zapatistas and underscored by a number of non-Zapatista communities is the issue of autonomy. In fact, in one of the rare references to natural resources in the San Andrés Accords, in January 1997 the Zapatistas demanded "the collective access to the use and enjoyment of the natural resources on their lands and territories" (Zapatista Communiqués Web Page). Likewise, the Comunidad Lacandona has appealed to the international donor community, particularly the World Bank, to allow local communities to be direct managers of their socioeconomic development and the conservation of their natural resources. They argue that the resulting proposals would reflect the realities and priorities of the communities, rather than those of the government or conservation sectors (Balboa 1996).

Indeed, the realities and priorities of the communities have become all the more pressing as traditional security nets continue to be dismantled, and Mexico slowly recovers from yet another financial crisis. However, as Leyva Solano and Ascencio Franco (1996:147, own translation) note, the colonizers of the Selva Lacandona are not sitting around waiting for the solution to their development problems to arrive:

> They themselves are conscious of the need to intensify the use of existing clearings; they have confronted the problem of deforestation and agreed within their ejidos to cut no more forest for pasture; they have declared zones of communal reserves with the objective of conserving the 'house of the howler monkey' and benefitting future generations. Through their organizations they have prepared concrete proposals for improving their standard of living and social well-being.

Conceding autonomy to communities in the conservation of tropical forests may be one means of finding common ground between social and environmental struggles in Chiapas. Moving the debates and discussions out of conservation circles and into community-based forums would allow individuals and groups to have a larger voice in the future of the Selva Lacandona. Such an approach supports the conclusions of Silva (1994:701), who found that "projects and policies work best when small rural communities participate with strong autonomous organization in planning and execution, rather than simply as the managers of projects designed by first world specialists."

Clearly not all communities will develop or support successful conservation strategies. For one thing, indigenous communities and mestizos in the Selva Lacandona have very different relationships with the natural environment (Arizpe et al. 1996). However, given the precarious status of the region's remaining forests and the limited amount of resources directed at the area, it is perhaps more desirable to work with the communities that are interested in developing conservation strategies, rather than with those that are resistant. It is possible that positive examples of community-led conservation projects will influence the perceptions of resistant communities, eventually generating ideas or initiatives that may take hold.

Environmentalists, conservation groups and government authorities can play an important role in facilitating local conservation plans or providing other types of support to communities. As Silva (1994:716) notes, "[i]nternational actors and domestic NGOs can provide critical support for local communities in their efforts to shape policies emanating from relatively sympathetic government offices." In fact, this is precisely the direction that these groups have been moving toward in recent years. For example, the recently established Consejo Técnico Asesor for the Montes Azules Biosphere Reserve offers the potential for opening up the management of protected areas to a much broader constituency. Likewise, Mexico's regional sustainable development programs (PRODERS) can create opportunities for local voices to be heard regarding both conservation and development. Nevertheless, given the existing tensions between environmental and social struggles in the Selva Lacandona region, it is essential that some conservation strategies originate at the local level, and that institutional and financial support are forthcoming to facilitate the strategies.

Providing groups with the explicit autonomy of deciding their own future and that of the environment around them could change patterns of deforestation in the years to come. However, the future of the region's tropical forest ecosystems, particularly the land protected within biosphere reserves, will largely rest with political decisions regarding continued militarization of the region; the outcome of the Zapatista conflict; the strength and persistence of the coalition of the Comunidad Lacandona; the resolution of agrarian conflicts throughout Chiapas, including contradictory titles in the Selva Lacandona; the successful incorporation of local people in management plans; the development of community-based forestry; and the future role of peasants in the global economy.

The word "sacrifice" suggests giving up or destroying something for an ideal, belief or end. Considering the driving forces of deforestation presented in this book, and the prominent distinction between environmental and social struggles evident in the region today, it becomes quite clear that deforestation indeed sustains a number of ideals, supports an assortment of beliefs, and serves a multitude of ends. The future of tropical forests is not guaranteed through social changes, just as social changes are not guaranteed through tropical forest conservation. The challenge in the Selva Lacandona is to overcome this polarizing distinction and recognize the mutual importance of environmental and social struggles in Chiapas. Until the distinction is dissolved, the Selva Lacandona will continue to be sacrificed to the logic of social, economic, and political relations in a rapidly changing world.

Notes

Chapter 1

1. Of a total area of approximately 1.9 million hectares, 1.75 million hectares of the Selva Lacandona region are located in the municipios of Ocosingo, Las Margaritas, and Altamirano. The other 150,000 hectares are found in Palenque, Chilón, La Trinitaria and La Independencia.

Chapter 2

1. *Ejidos* represent a system of communal land tenure institutionalized under post-revolutionary agrarian reforms after 1915. Members of ejidos, known as *ejidatarios*, hold usufruct land rights, either as individuals or collectively.

Chapter 3

1. There is an extensive literature on the Maya civilization, including works by Brainerd (1954), Thompson (1970) and Coe (1984).

2. Translation: Unknown desert inhabited by Lacandón Indians.

3. Ciudad Real is the historical name for the city now known as San Cristóbal de las Casas. The city's name was changed in 1848, to honor the first bishop of Chiapas, Bartolomé de las Casas.

4. The rough translation is "where there's a will, there's a way."

5. Ocosingo, the biggest municipio in Chiapas, contains the largest portion of the Selva Lacandona. In the course of negotiations between the government and the Zapatistas, the division of Ocosingo and Las Margaritas into smaller, more autonomous administrative units has formed an important basis for discussions.

6. Landsat Multispectral Scanner (MSS) and Thematic Mapper (TM) images were classified using a multicluster blocks technique to prepare a training set for a supervised maximum likelihood classification (see Hoffer 1986). A first-order classification was performed to distinguish between forest or natural vegetation cover and cleared areas.

7. *Milpa* is an Indian word used to refer to a cultivated plot. It is generally part of the slash-and-burn system of rotational cultivation.

8. As a result of the extensive bureaucracy involved in resolving land tenure issues, the resolutions were not officially executed until 1979 for Benemérito and 1980 for Quiringuicharo.

Chapter 4

1. Chiclé, the latex from the chicozapote tree (*Manilkara zapota L*.), was collected for use in chewing gum, as well as for industrial purposes during the Second World War. The Selva Lacandona served as a rich source of chiclé for companies such as Adams Chewing Gum, Corp. and Wrigley's, Inc. of the United States. An estimated 2,500 to 3,000 chicleros were distributed throughout twenty camps in the Selva Lacandona, many of which could only be reached by plane. Unlike the loggers, chicleros seem to have been treated well and justly remunerated for their labor (Moscoso Pastrana 1966). The chiclé industry declined in importance after World War II, when synthetic rubbers replaced the natural latex. Although a co-op was organized in 1973 for chicle collection in Ocosingo, it played only a minor role in the regional economy (Burguete Cal y Mayor 1978; Price and Hall 1983).

2. The myth of the pristine tropical rain forest of Latin America is addressed by Denevan (1992), who argues that human presence in the Americas was substantially larger at the time of colonization than three centuries later. Consequently, many of the forests that appear to be pristine are, in fact, anthropogenic in form and composition.

3. The head office was usually located in San Juan Bautista, Tabasco. There was also a branch office in Tenosique, and representatives stationed at shipping ports along the Gulf of Mexico (de Vos 1988a).

4. Stories of the horrors experienced by loggers emerged from the forest, either through witnesses or actual employees, such as Rubén Navarro who worked in the logging camps of Casa Bulnes Hermanos. The stories were then widely disseminated, largely through the novels of German author B. Traven (written between 1931 and 1940) and Pablo Montañez (written in the 1960s and 1970s). Solid documentation concerning the operation of the logging camps is scarce, both because the camps were geographically isolated, and because a low percentage of the workers survived the harsh work conditions (Benjamin 1981).

5. The literal translation is "hookers."

6. The practice was strongly supported in Chiapas, and in fact publicly defended by some as "a humane, efficient and legal contractual arrangement" (Benjamin, 1989:59). It was not until communal village lands were broken up in the 1890s, freeing up a new labor supply, that opposition to indebted servitude emerged within Chiapas. Initiatives to change the system were largely framed in terms of improving agricultural productivity, freeing up the pesos tied up in debt, and addressing the distribution of labor within the state. Nonetheless, reforms were continuously defeated in the state legislature until the late 1920s, when they were adopted without controversy.

7. The mahogany and tropical cedar trees extracted from the Selva Lacandona were exported as raw trunks to Europe and North America, where they would be manufactured into ships, furniture, and other goods. As a raw material, the prices for the wood were low in comparison with the value of the finished goods. González Pacheco (1983) attributes the lack of processing or manufacturing of hardwoods within Mexico to the European and North American

importers, who as primary sources of capital for the logging industry were simply not interested in developing local competition in manufactured goods.

8. This is not to suggest that modernization was not attempted by at least some of the companies. For example, in the post-World War I period, Agua Azul Co. introduced tractors to their logging camps along the Río Usumacinta. Once the difficulty of transporting the machines into the forest had been overcome, the company faced the logistical problems and expenses of getting fuel supplies to the camp. The only way to get gasoline was to transport it in wooden crates by mule from Tenosique, Tabasco (González Pacheco 1983). Consequently, the tractors remained idle and soon deteriorated in the tropical environment.

9. The total land grant was increased to 662,000 hectares in September of 1975, in order to include the Lacandones settled in Mensabok and Naja within the titled territory (Nations 1979).

10. Researchers such as Nations (1979) and de Vos (1980) argue that the present-day Lacandones are not the same as those which originally inhabited the forest prior to the Spanish Conquest. Instead, their arrival has been traced to the eighteenth century, from the Yucatan Peninsula. According to this research, the Lacandones *have not* been in possession of the Selva Lacandona "since time immemorial."

11. International mahogany prices did not play as important a role after 1949, as most of the timber was directed toward the domestic market.

12. SEMARNAP is the Secretaría de Medio Ambiente, Recursos Naturales y Pesca, Mexico's environmental ministry.

13. *Sexenio* refers to the six-year presidential term in Mexico.

Chapter 5

1. This interpretation is in contrast to the analysis of Barry (1995:105), who argues that the most directly affected are "the medium- and large-scale producers whose productivity permitted them to profit under a system of guarantee prices and subsidized inputs. The impact of falling prices on the production decisions of subsistence and infrasubsistence is not as direct, since this sector either does not market its produce or has never benefited from guarantee prices because of its remote locations and the corresponding control by private intermediaries over grain purchases and sales."

2. The fact that Chiapas is one of the principal producers of coffee, cacao, bananas and cattle is not coincidental, but rather it emanates from national policies to promoting these crops, supported by landowners interested in conserving and in many cases amplifying their property holdings. The only way to achieve this was to orient production towards these crops and cattle (Reyes Ramos 1992).

3. In terms of the Spanish colonial empire, Chiapas maintained a subordinate position because it lacked mineral resources that could add to the empire's wealth. Instead, Chiapas was valued largely as a source of agricultural products, including cacao, cotton, cochineal, indigo, dyewoods, and furs (Muench 1982).

4. Expropriation of Indian lands was one way of creating a wage labor force capable of fulfilling the needs of a growing mercantile economy. It also fulfilled the more immediate objectives of 1) putting the largest amount of productive land under cultivation; 2) stimulating land accumulation among the ladino population, who were thought to be more "enlightened" than the Indians, widely considered to be ignorant and lazy; and 3) integrating large properties which would return rich profits to the regional government through property tax collections (Marion Singer 1988:38).

5. In 1900, Chiapas governor Rafael Pimentel granted coffee growers in Soconusco the right to employ Highland Indians on their plantations. Prior to that, most labor had been brought in from neighboring Guatemala. Although a similar system of *enganche* was used for the coffee plantations as in the logging industry, most Indians voluntarily signed on to work during the coffee harvests because the wages were much better than in the Central Highlands: "Population increase and the diminution of communal lands in the highlands created the economic necessity of yearly migration to the harvests" (Benjamin, 1989:77).

Chapter 6

1. The 1923 *Ley de Tierras Libres* gave any Mexican above the age of 18 the right to acquire national lands or unused lands simply by notifying the *Secretaría de Agricultura y Fomento* that the parcels would be occupied (Reyes Ramos, 1992).

2. In practice, most of the NCPEs chose to distribute all 50 hectares to individual members, rather than cultivate them collectively (Neubauer 1997).

3. Many of the colonizers who left the *fincas* maintained the name in their new colony, enabling colonization patterns to be traced. For example, the workers from El Momón founded Nuevo Momón; those from Las Delicias founded Delicias Casco and San Antonio Las Delicias, and so on (Ascencio Franco and Leyva Solano 1992).

4. Construction of the road into the forest did not begin until 1970, but its purpose was evidently not to connect the ejidos to regional markets, but rather to establish a presence along the Guatemalan border (Garza Caligaris and Paz Salinas 1986).

Chapter 7

1. Translation published in the August 3, 1994 edition of the Anderson Valley Advertiser, Boonesville, CA. This special edition included English translations of all of the Zapatista communiqués through June, 1994.

2. In the course of dam construction during the 1970s, large amounts of agricultural lands were flooded. La Angostura project affected 604 km2; the Malpaso project affected 294 km2; and the Chicoasén project flooded 31.5 km 2.

3. Data on the number of presidential resolutions are often cited as evidence of progress in land distributions. However, resolutions do not imply that lands were actually turned over to campesinos — this depends instead on the execution. The *Ley Federal de Reforma Agraria*, Article 51, established that

from the time of publication in the *Diario Oficial de la Federación*, the nucleus of the ejidal population owns the land indicated in the resolution. Yet the process does not formally conclude until the execution of a presidential resolution which authorizes or confirms the ownership by the ejido in possession of the lands. The law does not establish a time frame under which this must occur. As a result, in many cases the execution of a presidential resolution takes as much as 40 years (Reyes Ramos 1992). The delay in officially transferring land to campesinos is largely a result of excessive bureaucratization: From the time of receipt of the initial land request to the time it is executed, Reyes Ramos (1992) calculates that 32 documents and 22 different agencies are involved. In Chiapas, it takes an average of 7.36 years to complete the procedures.

4. The highway connected with the existing road from PEMEX's Campo Lacantún to Palenque. Although it was originally designed to cross over the Ríos Chajul and Ixcán and join with the road to the Lagos de Montebello and Comitán, the portion from Chajul to Ixcán was incomplete as of 1996.

Chapter 8

1. Concern over the loss of forests emerged as early as the second half of the nineteenth century in Mexico. In 1881, a forest conservation law was promulgated which included an article requiring that ten seeds of mahogany, tropical cedar, or other species be planted for every single tree cut down (de Vos, 1988). However, this law was rarely complied with.

Chapter 9

1. The Selva Lacandona is not an isolated example of tensions between social and environmental struggles. In Guatemala, thirteen Conservation International employees were taken hostage at gunpoint in March, 1997. Sixty men ambushed the field station and burned it down, demanding that CI withdraw from the Laguna del Tigre part so that they could colonize the region. Subsequent negotiations paved the way for collaborations between the settlers and CI (Conservation International Web Page).

References

Abrams, Elliot M., AnnCorinne Freter, David J. Rue, and John D. Wingard. 1996. "The Role of Deforestation in the Collapse of the Late Classic Copan Maya State." Pages 55-75 in L. E. Sponsel, T. N. Headland, and R. C. Bailey, eds., *Tropical Deforestation: The Human Dimension*. New York: Columbia University Press.

Agrupación Sierra Madre, S.C. No date. *La Selva Lacandona: Su Conservación, una Prioridad*.

Allen, Julia C., and Douglas F. Barnes. 1985. "The Causes of Deforestation in Developing Countries." *Annals of the Association of American Geographers*, 75(2): 163-184.

Alvarez del Toro, Miguel. 1985. *¡Asi Era Chiapas!* Tuxtla Gutierrez, Chiapas, Mexico: Fundamat.

Anonymous. 1980. "Ecology Law in Mexico." *Nature and Resources*, 16(4):32.

Arizpe, Lourdes. 1996. *Sustentabilidad en las Áreas Selváticas: Equilibrar Derechos y Oportunidades*. Cuernavaca: UNAM, Centro Regional de Investigaciones Multidisciplinarias.

Arizpe, Lourdes, Fernanda Paz, and Margarita Velazquez. 1996. *Culture and Global Change: Social Perceptions of Deforestation in the Lacandona Rain Forest in Mexico*. Ann Arbor: The University of Michigan Press.

Arizpe, Lourdes, M. Priscilla Stone, and David C. Major, eds. 1994. *Population and the Environment: Rethinking the Debate*. Boulder: Westview Press.

Ascencio Cedillo, Efraín. 1995. "Un Acercamiento Socio-histórico a la Ganadería de Ocosingo, Chiapas." Pages 75 - 124 in *Anuario 1995*. Tuxtla Gutiérrez, Mexico: Centro de Estudios Superiores de México y Centroamérica, Universidad de Ciencias y Artes del Estado de Chiapas.

Ascencio Franco, Gabriel, and Xochitl Leyva Solano. 1992. "Los Municipios de la Selva Chiapaneca." Colonización y Dinámica Agropecuaria. Pages 176-241 in Anuario del Instituto Chiapaneco de Cultura. Tuxtla Gutiérrez, Chiapas, Mexico: Instituto Chiapaneco de Cultura.

Atkinson, Adrian. 1991. *Principles of Political Ecology*. London: Bellhaven Press.

Balboa, Juan. 1996. "Congeló el banco 20 mdd ante diferencias en criterios de aplicación." *La Jornada en Internet*, 5 April 1996.

_____. 1997. "Prolifera la prostitución en zonas militares de Chiapas." *La Jornada en Internet*, 27 January 1997.

Ballinas, Juan. 1951. *El Desierto de Los Lacandones: Memorias, 1876-1877*. Tuxtla Gutiérrez, Chiapas, Mexico: Publicaciones del Ateneo de Chiapas.

Barkin, David. 1990. *Distorted Development: Mexico in the World Economy*. Boulder: Westview Press.

Barkin, David, and Blanca Suarez. 1985. *El fin de la autosuficiencia alimentaria*. Mexico: Ecodesarrollo/Ediciones Oceano, S.A.

Barry, Tom. 1995. *Zapata's Revenge: Free Trade and the Farm Crisis in Mexico*. Boston: South End Press.

Bartra, Roger. 1993. *Agrarian Structure and Political Power in Mexico*. Baltimore: The Johns Hopkins University Press.

Benjamin, Thomas. 1981. El Trabajo en las Monterías de Chiapas y Tabasco, 1870-1946. *Historia Mexicana*, 30(4): 506-529.

_____. 1989. *A Rich Land, A Poor People: Politics and Society in Modern Chiapas*. Albuquerque: University of New Mexico Press.

Blaikie, Piers, and Harold Brookfield. 1987. *Land Degradation and Society*. London: Methuen.

Boo, Elizabeth. 1990. *Ecoturismo: Potenciales y Escollos*. Washington: World Wildlife Fund and The Conservation Foundation.

Brainerd, George W. 1954. *The Maya Civilization*. Los Angeles: Southwest Museum.

Bray, David Barton. 1997. *Forest and Protected Areas Policies in the Lacandon Rainforest, Chiapas*. Paper presented at the XX International Congress of the Latin American Studies Association, April 17 - 19, 1997 in Guadalajara, Mexico.

Bryant, Raymond L. 1992. Political Ecology: An Emerging Research Agenda in Third-World Studies. *Political Geography*, 11(1):12-36.

Buerkle, Dorothee. 1996. *Die Landnutzungsveränderung im Lacandona-Regenwald (Chiapas, Mexiko) zwischen 1974 und 1993 untersucht mit multitemporalen Landsat MSS und TM Daten*. Unpublished Diplomarbeit, Philipps-Universität Marburg. Marburg, Germany.

Bunker, Stephen G. 1985. *Underdeveloping the Amazon: Extraction, Unequal Exchange, and the Failure of the Modern State*. Chicago: University of Illinois Press.

Burguete Cal y Mayor, Aracely. 1978. "La Selva Lacandona: ¿Desarrollo o Crecimiento?" Pages 29-67 in *Indigenismo: Evaluación de una Práctica*. Mexico City: Instituto Nacional Indígenista.

Calleros, G., and F. A. Brauer. 1983. *Problematica regional de la Selva Lacandona*. Palenque, Chiapas: Direccion General de Desarrollo Forestal, Secretaría de Agricultura y Recursos Hidraulicos. Coordinación Ejecutiva del Programa Ecológico de la Selva Lacandona.

Carlsen, Laura. 1996. "Mexico's New Indian Movement." *Interhemispheric Resource Center Bulletin*, 45: 1-8.

Carmona Lara, Ma. Del Carmen. 1988. *Ecologia - Cambio Estructural en Chiapas: Avances y Perspectivas*. Tuxtla Gutierrez: Universidad Autonoma de Chiapas.

Casco Montoya, Rosario. 1990. "El Uso de los Recursos del Trópico Mexicano: El Caso de la Selva Lacandona." Pages 115-148 in E. Leff, ed., *Medio Ambiente y Desarrollo en México*. Mexico: Miguel Angel Porrua.

Castellanos Cambranes, Julio. 1984. "Origins of the Crisis of the Established Order in Guatemala." Pages 119-152 in S. C. Ropp and J. A. Morris, eds., *Central America: Crisis and Adaptation*. Albuquerque: University of New Mexico Press.

Castro Soto, Gustavo Enrique. 1994. "Guatemala y sus Refugiados en el Conflicto Armado de Chiapas." Pages 179-188 in M. B. Monroy, ed., *Pensar Chiapas, Repensar Mexico: Reflexiones de las ONGs Mexicanas sobre el Conflicto*. Mexico: Convergencia de Organismos Civiles por la Democracia.

Cockcroft, James D. 1990. *Mexico: Class Formation, Capital Accumulation, and the State*. New York: Monthly Review Press.

Coe, Michael D. 1984. *The Maya*. London: Thames and Hudson.

Collier, George A., Daniel C. Mountjoy, and Ronald B. Nigh. 1994. "Peasant Agriculture and Global Change." *BioScience*, 44(6):398-407.

Collier, George A., and Elizabeth Lowery Quaratiello. 1994. *BASTA! Land and the Zapatista Rebellion in Chiapas*. Oakland, CA: Food First.

Conpaz et al. 1996. *Militarización y Violencia en Chiapas*. Report by Coordinación de Organismos no Gubernamentales por la Paz (Conpaz), Centro de Derechos Humanos Fray Bartolome de Las Casas, and Convergencia de Organismos Civiles por la Democracia. Mexico City: Servicios Informativos Procesados (SIPRO).

Cortez-Ortiz, A. 1990. *Estudio Preliminar sobre Desforestación en la Región Fronteriza del Río Usumacinta*. Internal Report. Mexico: INEGI.

Cruz, Angeles. 1997. "Se han perdido 100 mil hectáreas de bosques en Chiapas desde 94: Quadri." *La Jornada en Internet*, 21 June 1997.

Cruz Coutiño, J. Antonio and Blanca Esthela Parra Chávez. 1994. *Política Forestal de Chiapas, Antecedentes y Expectativas*. Unpublished Thesis. Licenciado en Sociologia y Economia, Universidad Autonoma de Chiapas, Mexico.

de Janvry, Alain. 1981. *The Agrarian Question and Reformism in Latin America*. Baltimore: The Johns Hopkins University Press.

de Janvry, Alain, Gustavo Gordillo, and Elisabeth Sadoulet. 1997. *Mexico's Second Agrarian Reform: Household and Community Responses*. San Diego: Center for U.S.-Mexico Studies, UCSD.

de Vos, Jan. 1980. *La Paz de Dios y del Rey: La Conquista de la Selva Lacandona, 1525-1821*. Mexico City: Fondo de Cultura Económica.

_____. 1988a. *Oro Verde: La Conquista de la Selva Lacandona por los Madereros Tabasqueños, 1822-1949*. Mexico City: Fondo de Cultura Económica.

_____. 1988b. *Viajes al Desierto de la Soledad: Cuando la Selva Lacandona aún era Selva*. Mexico City: Secretaría de Educación Pública.

_____. 1992. "Una Selva Herida de Muerte, Historia Reciente de la Selva Lacandona." Pages 267-286 in M. A. Vásquez Sánchez and M. A. Ramos, eds., *Reserva de la Biósfera Montes Azules, Selva Lacandona: Investigación para su Conservación*. San Cristóbal de las Casas, Chiapas: Publ. Esp. Ecosfera.

Denevan, William M. 1992. "The Pristine Myth: The Landscape of the Americas in 1492." *Annals of the Association of American Geographers*, 82(3):369-385.

Detwiler, R. P., and C.A.S. Hall. 1988. Tropical Forests and the Global Carbon Cycle. *Science* 239:42-47.

Diario Oficial. 1972. Decreto mediante el cual se titulan 614,321 ha a favor de 66 jefes de familia lacandones. 6 March 1972.

_____. 1978. "Decreto por el que se declara de interés publico el establecimiento de la zona de protección forestal de la cuenca del Rio Tulijah, así como de la reserva integral de la biósfera Montes Azules, en el area comprendida dentro de los limítes que se indican." 12 January 1978, Pages 6-8.

_____. 1992a. "Decreto por el que se declara área natural protegida con el carácter de area de flora y fauna silvestres la región Chan-Kin, con superficie de 12,184-98-75 hectáreas, ubicada en el Municipio de Ocosingo, Chis. (Segunda publicación). 24 August 1992, Pages 16-20.

_____. 1992b. "Decreto por el que se declara área natural protegida con el carácter de Monumento Natural, a la zona conocida con el nombre de Yaxchilán, con una superficie de 2,621-25-23 hectáreas, ubicada en el Municipio de Ocosingo, Chis." (Segunda publicacion). 24 August 1992, Pages 27-30.

_____. 1992c. "Decreto por el que se declara área natural protegida con el carácter de Monumento Natural, la zona conocida como Bonampak, con superficie de 4,357-40-00 hectáreas, ubicada en el Municipio de Ocosingo, Chis." (Segunda publicacion). 24 August 1992, Pages 20-22.

_____. 1992d. "Decreto por el que se declara área natural protegida con el carácter de Reserva de la Biósfera la zona conocida como Lacan-Tun, con un superficie de 61,873-96-02.5 hectáreas, ubicada en el Municipio de Ocosingo, Chis." (Segunda publicacion). 24 August 1992, Pages 22-26.

Dichtl, Sigrid. 1987. *Cae una Estrella: Desarrollo y Destrucción de la Selva Lacandona*. Mexico City: Secretaría de Educación Pública.

Dziedzic, Michael J. and Stephen J. Wager. 1992. "Mexico's Uncertain Quest for a Strategy to Secure its Southern Border." *Journal of Borderland Studies*, 7(1):19-48.

Ecosur. 1997. *SELVANET: Sistema de Información para el Desarrollo Sustentable de la Region Selva Lacandona, Chiapas*. Report to the Secretaría de Medio

Ambiente, Recursos Naturales y Pesca (SEMARNAP). San Cristóbal de las Casas: El Colegio de la Frontera Sur. (Ecosur).

El Día. 1987. "Grupos de cazadores y compañías madereras son los principales enemigos de la Selva Lacandona." *El Día.* 9 April 1987.

Elvira Vargas, Rosa. 1994. "Revisará la Sedeso proyectos y obras en la selva chiapaneca." *La Jornada.* 4 January 1994.

Ewel, John J. 1986. "Designing Agricultural Ecosystems for the Humid Tropics." *Annual Review of Ecology and Systematics,* 17:245-271.

Faber, Daniel. 1993. *Environment Under Fire: Imperialism and the Ecological Crisis in Central America.* New York: Monthly Review Press.

Falla, Ricardo. 1994. *Massacres in the Jungle: Ixcán, Guatemala (1975-1982).* Boulder: Westview Press.

Fernández Ortiz, Luis M., María del Carmen García Aguilar, María Tarrío García, Daniel Villafuerte Solís, amd Fermamdp Díaz Pérez. 1994. "Ganaderia, Deforestación y Conflictos Agrarios en Chiapas." *Cuadernos Agrarios,* Pages 20-48.

Fideicomiso de la Selva Lacandona. 1977. *Reporte Final de Actividades: Agosto de 1976 - Agosto de 1977* : Fideicomiso de la Selva Lacandona, San Cristóbal de las Casas, Chiapas, Mexico.

Floyd, J. Charlene. 1994. *The Catholic Church and Democratization in Contemporary Mexico: A Case Study of Catequiestas in Chiapas - some preliminary observations.* Paper presented at the Society for the Scientific Study of Religion and Religious Research Association, Albuquerque, NM.

Fuentes Aguilar, Luis and Consuelo Soto Mora. 1992. "Colonización y Deterioro de la Selva Lacandona." *Revista Geografica,* 116:67-84.

García Aguilar, María del Carmen and Daniel Villafuerte Solís. 1996. "A Propósito de la insurrección Zapatista. Notas Sobre Economía y Sociedad en Chiapas, Mexico." Pages 36 - 74 in *Anuario 1995.* Tuxtla Gutiérrez, Mexico: Centro de Estudios Superiores de México y Centroamérica, Universidad de Ciencias y Artes del Estado de Chiapas.

García de León, Antonio. 1984a. *Resistencia y Utopia, Tomo 1.* Mexico City: Ediciones Era.

_____. 1984b. *Resistencia y Utopia, Tomo 2.* Mexico City: Ediciones Era.

_____. 1985. "De Mozos, Hierros y Ganados: La Ganaderia Chiapaneca como una Permanencia Conflictiva." *Ensayos: Economía, Politica e Historia,* 2(7): 42-59.

García de León, Antonio, and Carlos Monsivais. 1995. *EZLN: Documentos y Comunicados 2, 15 de Agosto de 1994 / 29 de Septiembre de 1995.* Mexico: Ediciones Era.

García de León, Antonio, Elena Poniatowska, and Carlos Monsivais. 1994. *EZLN: Documentos y Comunicados, 1 de Enero / 8 de Agosto de 1994.* Mexico: Ediciones Era.

Garza Caligaris, Anna María and María Fernanda Paz Salinas. 1986. "Las Migraciones: Testimonios de Una Historia Viva." Pages 89-105 in *Anuario,* Vol. 1. San Cristóbal de las Casas, Chiapas: Centro de Estudios Indigenas/Universidad Autonoma de Chiapas.

GEF. 1992. Protected Areas in Mexico. Global Environmental Facility Report. Washington: World Bank.

Gillis, Malcolm, and Robert Repetto. 1988. "Conclusions: Findings and Policy Implications." Pages 385-410 in R. Repetto and M. Gillis, eds., *Public Policies and the Misuse of Forest Resources.* New York: Cambridge University Press.

Gobierno del Estado de Chiapas. 1988. *Los Municipios de Chiapas.* Mexico: Colección: Enciclopedia de los Municipios de México.

Godinez Herrera, José Tomás. 1989. *El Estado y la Explotación Maderera (Caso: La Comunidad Lacandona).* Unpublished Thesis, Universidad Autonoma Chapingo, Chapingo.

Gómez-Pompa, Arturo. 1992a. "Una Visión Sobre el Manejo del Tropico Humedo de Mexico." Pages 7-18 in M. A. Vásquez Sánchez and M. A. Ramos, eds., *Reserva de la Biósfera Montes Azules, Selva Lacandona: Investigacion para su Conservación*. San Cristóbal de las Casas, Chiapas: Publ. Esp. Ecosfera.

_____. 1992b. *Yaxbe: Camino Verde*. Propuesta para ampliar el area protegida de la Selva Lacandona y establecer un corredor biologico con las selvas del "gran Peten." Unpublished document.

Gómez-Pompa, A., and J. Valdes. 1962. "Una nueva especie de Yucca de la Selva Lacandona." *Boletín de la Sociedad Botanico de Mexico*, 27:3-46.

Gómez-Pompa, A., C. Vázquez-Yanes, and S. Guevara. 1972. "The Tropical Rain Forest: A Nonrenewable Resource." *Science*, 177(4051):762-765.

González Pacheco, Cuauhtémoc. 1983. *Capital Extranjero en la Selva de Chiapas: 1863 - 1982*. Mexico City: Instituto de Investigaciones Económicas, UNAM.

González-Garcia, Fernando. 1992. "Aves de la Selva Lacandona, Chiapas, Mexico." Pages 173-200 in M. A. Vásquez Sánchez and M. A. Ramos, eds., *Reserva de la Biósfera Montes Azules, Selva Lacandona: Investigación para su Conservación*. San Cristóbal de las Casas, Chiapas: Publ. Esp. Ecosfera.

González Ponciano, Jorge Ramón. 1990. "Frontera, Ecología y Soberanía Nacional: La Colonización de la Franja Fronteriza Sur de Marqués de Comillas." Pages 50-83 in *Anuario 1990*. Tuxtla Gutiérrez, Chiapas: Instituto Chiapaneco de Cultura.

Grainger, Alan. 1993. "Rates of Deforestation in the Humid Tropics: Estimates and Measurements." *The Geographical Journal*, 159:33-44.

Greenberg, James B., and Thomas K. Park. 1994. Political Ecology. *Journal of Political Ecology*, 1 (http://www.library.arizona.edu/ej/jpe/volume_1).

Greenberg, Russell. 1990. *Southern Mexico: Crossroads for Migratory Birds*. Washington: Smithsonian Migratory Bird Center.

Gregg, William, and Betsy McGean. 1985. "Biosphere Reserves." *Orion Nature Quarterly*, 4(3):41-51.

Grindle, Merilee S. 1986. *State and Countryside: Development Policy and Agrarian Politics in Latin America*. Baltimore: The Johns Hopkins University Press.

Gutelman, Michel. 1971. *Capitalismo y Reforma Agraria en México*. Mexico City: Ediciones Era.

Halffter, Gonzalo. 1980a. "Biosphere Reserves and National Parks: Complementary Systems of Natural Protection." *Impact of Science on Society*, 30(4):269-277.

_____. 1980b. *Biosphere Reserves: a New Method of Nature Protection*. Paper presented at the International Seminar on Social and Environmental Consequences of Natural Resource Policies, with Special Emphasis on Biosphere Reserves. Durango, Mexico.

_____. 1984. "Conservation, Development and Local Participation." Pages 428-436 in F. DiCastri, F. W. G. Baker, and M. Hadley, eds., *Ecology in Practice, Part I: Ecosystem Management*. Paris: UNESCO.

Halhead, Vanessa. 1984. *The Forests of Mexico: The Resource and the Politics of Utilization*. Unpublished Master of Philosophy, University of Edinburgh.

_____. 1992. "Social Dimensions of Forest Utilization in Mexico: Implications for Intervention." Pages 159-169 in T. E. Downing, S. B. Hecht, H. A. Pearson, and C. Garcia-Downing, eds., *Development or Destruction: The Conversion of Tropical Forest to Pasture in Latin America*. Boulder: Westview Press.

Harvey, Neil. 1995. "Rebellion in Chiapas: Rural Reforms and Popular Struggle." *Third World Quarterly*, 16(1):39-73.

_____. 1996. "Rural Reforms and the Zapatista Rebellion: Chiapas, 1988-1995." Pages 187-208 in G. Otero, ed., *Neo-Liberalism Revisited: Economic Restructuring and Mexico's Political Future*. Boulder: Westview Press.

_____. 1997. *Economic Viability on the Post-Modern Frontier: Illegal Logging, the Ejido and the State in Marques de Comillas, Chiapas.* Paper presented at the XX International Congress of the Latin American Studies Association, April 17-19 in Guadalajara, Mexico.

Hecht, Susanna, and Alexander Cockburn. 1989. *The Fate of the Forest*. New York: Verso.

Hernández Castillo, Rosalva Aída. 1992. "Los Refugiados Guatemaltecos y la Dinámica Fronteriza en Chiapas." Pages 93-105 in G. Freyermuth Enciso and R. A. Hernández Castillo, eds., *Una Década de Refugio en México*. Mexico: Centro de Investigaciones y Estudios Superiores en Antropología Social.

Hernández, Evangelina. 1990. "Destruido, el 76% de la Selva Lacandona." *La Jornada*, 21 July 1990.

Hernández Escobar, J. L. 1992. *Impacto Socioeconomico de PEMEX en el Distrito de Ocosingo*. Unpublished Thesis. Licenciado en Economia, Universidad Autonoma de Chiapas, San Cristóbal de las Casas.

Hewitt de Alcántara, Cynthia. 1973. "The 'Green Revolution' as History: the Mexican Experience." *Development and Change*, 15(2): 25-44.

_____. 1994. "Introduction: Economic Restructuring and Rural Subsistence in Mexico." Pages 1-24 in C. Hewitt de Alcantara, ed., *Economic Restructuring and Rural Subsistence in Mexico*. San Diego: Center for U.S.-Mexican Studies, UCSD.

Hoffer, Roger M. 1986. "Digital Analysis Techniques for Forestry Applications." *Remote Sensing Reviews*, 2:61-110.

Howard, Philip, and Thomas Homer-Dixon. 1996. *Environmental Scarcity and Violent Conflict: The Case of Chiapas, Mexico*. The Project on Environment, Population and Security. American Association for the Advancement of Science, University College at University of Toronto.

INE. 1997. *Ley Forestal de México*.

INEGI. 1990. *Anuario Estadistico del Estado de Chiapas*. Aguascalientes, Mexico: Instituto Nacional de Estadistica, Geografía e Informática.

_____. 1991. *Chiapas: XI Censo General de Poblacion y Vivienda, 1990: Resultados Definitivos - Datos por Localidad (Integracion Territorial) Tomo I.* Aguascalientes, Mexico: Instituto Nacional de Estadistica, Geografía e Informática.

_____. 1993. *Anuario Estadistico del Estado de Chiapas, Edicion 1993*. Aguascalientes, Mexico: Instituto Nacional de Estadistica, Geografía e Informática.

_____. 1994. *Censos Agrícola, Sector Agropecuario, Ganadero y Ejidal, 1991.* Aguascalientes, Mexico: Instituto Nacional de Estadistica, Geografía e Informática (CD-ROM).

_____. 1995. *Estadísticas del Medio Ambiente, México 1994*. Aguascalientes, Mexico: Instituto Nacional de Estadistica, Geografía e Informática.

Iñigo-Elías, Eduardo. 1991. *Ecological Correlates of Forest Fragmentation and Habitat Alteration on a Tropical Raptor Community in the Selva Lacandona Region of Chiapas, Mexico*. Unpublished Masters Thesis. University of Florida, Gainesville.

Katzenberger, Elaine, ed. 1995. *First World, ha ha ha! The Zapatista Challenge*. San Francisco: City Lights.

Lazcano-Barrero, Marco A., Eleuterio Gongora-Arones, and Richard C. Vogt. 1992. "Anfibios y Reptiles de la Selva Lacandona." Pages 145-171 in M. A. Vásquez Sánchez and M. A. Ramos, eds., *Reserva de la Biósfera Montes Azules, Selva Lacandona: Investigación para su Conservación*. San Cristóbal de las Casas, Chiapas: Publ. Esp. Ecosfera.

Lazcano-Barrero, Marco A., Ignacio J. March, and Miguel Angel Vásquez-Sánchez. 1992. "Importancia y Situacion Actual de la Selva Lacandona:

Perspectivas para su Conservación." Pages 393-436 in M. A. Vásquez Sánchez and M. A. Ramos, eds., Reserva de la Biósfera Montes Azules, Selva Lacandona: Investigación para su Conservación. México: Publicaciones Especiales Ecosfera.

Leyva Solano, Xochitl, and Gabriel Ascencio Franco. 1991. Espacio y Organizacion Social en la Selva Lacandona: El Caso de la Subregion Cañadas Pages 17-49 in *Anuario del Instituto Chiapaneco de Cultura*. Tuxtla Gutiérrez, Chiapas, México: Instituto Chiapaneco de Cultura.

_____. 1993. "Apuntes para el Estudio de la Ganaderización en la Selva Lacandona." Pages 262-284 in *Anuario del Instituto Chiapaneco de Cultura*. Tuxtla Gutierrez, Chiapas, Mexico: Instituto Chiapaneco de Cultura.

_____. 1996. *Lacandonia al Filo del Agua*. Mexico City: Fondo de Cultura Economica.

Lipietz, Alain. 1995. *Green Hopes: The Future of Political Ecology*. Cambridge: Polity Press.

Lobato, Rodolfo. 1980. "Estratificacion Social y Destrucción de la Selva Lacandona en Chiapas." *Revista Ciencia Forestal*, 5(24):45-54.

_____. 1981. "Antropologia Economica de las Comunidades Mayas de la Selva Lacandona, Chiapas." Pages 231-238 in *Investigaciones Recientes en el Area Maya: XVII Mesa Redonda*. Proceedings from the annual meeting of the Sociedad Mexicana de Antropología, 21-27 June 1981, San Cristóbal de las Casas, Chiapas.

López A., Martha Patricia. 1996. *La Guerra de Baja Intensidad en Mexico*. Mexico: Plaza y Valdes Editores.

MacIntyre, Loren. 1977. "Brazil's Wild Frontier." *National Geographic Magazine*, November: 684-719.

Manz, Beatriz. 1988. *Refugees of a Hidden War: The Aftermath of Counterinsurgency in Guatemala*: State University of New York Press.

March, Ignacio J. 1994. *Diagnostico Actualizado de la Reserva Integral de la Biósfera Montes Azules, Chiapas* . Chiapas, Mexico. Sub-estudio de Areas Protegidas de México, Estudio del Subsector Forestal del Banco Mundial. Unpublished report.

_____. 1997. *Consejo Técnico Asesor de la Reserva de la Biósfera Montes Azules, Selva Lacandona, Chiapas*. Unpublished report. El Colegio de la Frontera Sur.

March, Ignacio J., and Marcelo Aranda. 1992. Mamiferos de la Selva Lacandona, Chiapas 201-220 in M. A. Vásquez Sánchez and M. A. Ramos, eds., *Reserva de la Biósfera Montes Azules, Selva Lacandona: Investigación para su Conservación*. San Cristóbal de las Casas, Chiapas: Publ. Esp. Ecosfera.

March, I.J., Naranjo, E., Rodiles, R., Navarrete, D., Alba, M.P., Hernández, P. and V.H. Loaiza. 1996. *Diagnóstico para la Conservación y Manejo de la Fauna Silvestre en la Selva Lacandona, Chiapas*. Final report for the Subdelegación de Planeación in Chiapas, for the Secretaría de Medio Ambiente, Recursos Naturales y Pesca (SEMARNAP) - ECOSUR, San Cristóbal de las Casas, Chiapas. Unpublished.

Marion Singer, Marie-Odile. 1988. *El Agrarianismo en Chiapas (1524-1940)*. Mexico City: Instituto Nacional de Antropologia e Historia.

Martínez, E. and C. Ramos. 1989. "Lacandoniaceae (Triudirales): Una Nueva Familia de México." *Annals of the Missouri Botanical Garden*, 76:128-135.

Masera, Omar. 1996. *Desforestación y Degradación Forestal en México*. Report No. 19. Michoacán: Grupo Interdisciplinario de Tecnología Rural Apropiada (GIRA).

Mauricio Leguizamo, Juan M., Ruben Valladares Arjona, and Hector García Juaréz. 1985. *Lacandona: Una Incorporación Anarquica al Desarrollo Nacional*. Partido Revolucionario Institucional, Centro de Investigaciones Ecologicas del Sureste (CIES).

Medellín, Rodrigo A. 1991. The Selva Lacandona: An Overview. *TCD Newsletter* (Tropical Conservation and Development Program, University of Florida, Gainesville), 24:1-5.

_____. 1994. "Mammal Diversity and Conservation in the Selva Lacandona, Chiapas, Mexico." *Conservation Biology*, 8(3):780-799.

Medellín, Rodrigo A., Oscar Sánchez Herrera, and Guillermina Urbano V. 1992. Ubicacion zoogeografica de la Selva Lacandona, Chiapas, Mexico, a traves de su fauna de quiropteros 233-251 in M. A. Vásquez Sánchez and M. A. Ramos, eds., *Reserva de la Biósfera Montes Azules, Selva Lacandona: Investigación para su Conservación*. San Cristóbal de las Casas, Chiapas: Publ. Esp. Ecosfera.

Monroy, Mario B., ed. 1994. *Pensar Chiapas, Repensar Mexico: Reflexiones de las ONGs Mexicanas Sobre el Conflicto*. Mexico: Convergencia de Organismos Civiles por la Democracia.

Morón, Miguel Angel. 1992. Estado actual de conocimiento sobre los insectos de la Selva Lacandona 119-134 in M. A. Vásquez Sánchez and M. A. Ramos, eds., *Reserva de la Biósfera Montes Azules, Selva Lacandona: Investigación para su Conservación*. San Cristóbal de las Casas, Chiapas: Publ. Esp. Ecosfera.

Moscoso Pastrana, Prudencio. 1966. *La Tierra Lacandona*. San Cristo'bal de Las Casas, Chiapas: Corporacion de Fomento de Chiapas, S.A. de C.V.

Muench, Pablo E. 1982. Las Regiones Agrícolas de Chiapas. *Geografía Agrícola*, 2:57-102.

Myers, Norman. 1980. Deforestation in the Tropics: Who Gains, Who Loses? Pages 1-21 in V. H. Sutlive, N. Altshuler, and M. D. Zamora, eds., *Where Have All the Flowers Gone? Deforestation in the Third World* (Publication Number 13). Williamsburg, VA: Department of Anthropology, College of William and Mary.

_____. 1981. "The Hamburger Connection: How Central America's Forests Become North America's Hamburgers." *Ambio*, 10(1): 3-8.

_____. 1984. *The Primary Source: Tropical Forests and Our Future*. New York: W. W. Norton & Company.

_____. 1992. "Tropical Forests: The Policy Challenge." *The Environmentalist*, 12(1):15-27.

Nash, June. 1995 "The Reassertion of Indigenous Identity: Mayan Responses to State Intervention in Chiapas."*Latin American Research Review*, 30(3):7-41.

Nations, James D. 1979. *Population Ecology of the Lacandon Maya*. Unpublished Ph.D. Dissertation, Southern Methodist University.

_____. 1981. "The Rainforest Farmers." *Pacific Discovery*, 34(1):1-9.

_____. 1994. "The Ecology of the Zapatista Revolt." *Cultural Survival Quarterly*, Spring:31-33.

Nations, James D., and Ronald B. Nigh. 1980. "The Evolutionary Potential of Lacandon Maya Sustained-Yield Tropical Forest Agriculture." *Journal of Anthropological Research*, 36(1):1-30.

Nations, James D. and Daniel I. Komer. 1987. "Rainforests and the Hamburger Society." *The Ecologist*, 17(4/5):161-167.

Neubauer, Katrin. 1997. *Driving Forces of Tropical Deforestation In Mexico: The Ejido and Land Use Changes in Marques de Comillas, Chiapas*. Unpublished Masters Thesis, The University of Arizona.

O'Brien, Karen L. 1995. *Deforestation and Climate Change in the Selva Lacandona of Chiapas, Mexico*. Unpublished Ph.D. Dissertation, The Pennsylvania State University.

Ordóñez Díaz, María de Jesús and Oscar Flores Villela. 1995. *Áreas Naturales Protegidas*. Mexico: Pronatura.

Ovilla, Jose R. and Jorge A. Díaz López. 1985. "Devasta Pemex la Selva Lacandona; ante el ecocidio, interviene la Sedue y detiene tareas de exploración." *El Universal*. 3 April 1985.

Paz Salinas, María Fernanda. 1989. *La Migración a Las Margaritas: Una Historia a Dos Voces*. Unpublished Thesis. Licenciada en Antropologia Social, Universidad Autonoma de Chiapas.

_____. 1994. "La Selva Lacandona: De Tierra Prometida a Zona de Conflicto. Reflexiones Sobre El Futuro Sustentable en la Region." Pages 91-98 in D. Moctezuma Navarro, ed., *Chiapas: Los Problemas de Fondo*. Cuernavaca: UNAM, Centro Regional de Investigaciones Multidisciplinarias.

PEMEX. 1984. *Zona Lacandona: Gerencia de Desarrollo Regional*. Petroléos Mexicanos, Subdireccion de Planeacion y Coordinacion.

_____. 1986. *Proyecto: Desarrollo y Preservación de la Selva Lacandona (Diagnostico de las Areas con Posibilidades de Desarrollo Petrolero)*. Petroléos Mexicanos, Subdirección Tecnica Administrative, Coordinación Ejecutiva para el Desarrollo de las Zonas Petroleras.

Pérez Gil, Ramón. 1991. "Lacandonia: Controvertida y Amenazada." Pages 126-137 in S. C. Agrupacion Sierra Madre, ed., *Lacandonia: El Ultimo Refugio*. Mexico:Universidad Nacional Autonoma de Mexico.

Pólito, Elizabeth and Juan González Esponda. 1996. "Cronología. Veinte Años de Conflictos en el Campo: 1974-1993." *Chiapas* 2:197-220.

Preciado Llamas, Juan. 1976. "Una Colonia Tzeltal en la Selva Lacandona: Aspectos Socio-Económicos de su Relación con el Ecosistema." Pages 391-412 in R. Redfield and A. Villa R., eds., *Notes on the Ethnography of Tzeltal Communities of Chiapas*. Contributions to American Anthropology and History, No. 28.

_____. 1978. "Reflexiones Teorico-Metodológicas para el Estudio de la Colonización en Chiapas." Pages 45-67 in *Economía Campesina y Capitalismo Dependiente*. Mexico City: Universidad Nacional Autonoma de Mexico.

Price, Turner, and Lana Hall. 1983. *Agricultural Development in the Mexican Tropics: Alternatives for the Selva Lacandona Region of Chiapas*. Cornell/International Agricultural Economics Study, A.E. Research 83-4, Cornell University Department of Agricultural Economics.

Retiere, Alán. 1991. *¿El Ganado Contra la Selva?* Doc. 039-VI-91: Instituto de Asesoría Antropológica para la Región Maya, A.C.

Reyes Ramos, María Eugenia. 1992. *El Reparto de Tierras Y La Politica Agraria en Chiapas*. Mexico City: CIHMECH, UNAM.

Richards, J.F., and R.P. Tucker. 1988. "Introduction." Pages 1-12 in J. F. Richards and R. P. Tucker, eds., *World Deforestation in the Twentieth Century*. Durham: Duke University Press.

Rojas, Rosa. 1986. CIES: Pemex y CFE, responsables de la desaparición de la Selva Lacandona." *La Jornada*. 4 August 1986.

_____. 1995. *Chiapas: La Paz Violenta*. México: La`Jornada Ediciones.

Romero Jacobo, Cesár. 1994. *Los Altos de Chiapas: La Voz de las Armas*. México: Grupo Editorial Planeta.

Ross, John. 1995. *Rebellion from the Roots: Indian Uprising in Chiapas*. Monroe, ME: Common Courage Press.

Russell, P.L. 1995. *The Chiapas Rebellion*. Austin: Mexico Resource Center.

Sanderson, Steven E. 1986. *The Transformation of Mexican Agriculture: International Structure and the Politics of Rural Change*. Princeton: Princeton University Press.

Schmink, Marianne. 1994. "Socioeconomic Matrix of Deforestation." Pages 253-275 in L. Arizpe, M. P. Stone, and D. C. Major, eds., *Population and Environment: Rethinking the Debate*. Boulder: Westview Press.

Schmink, Marianne, and Charles H. Wood. 1987. "The 'Political Ecology' of Amazonia." Pages 38-57 in P. D. Little and M. M. Horowitz, eds., *Lands at Risk in the Third World*. Boulder: Westview Press.

Schwartz, Norman B. 1995. "Colonization, Development, and Deforestation in Peten, Northern Guatemala." Pages 101-130 in M. Painter and W. H. Durham, eds., *The Social Causes of Environmental Destruction in Latin America*. Ann Arbor: The University of Michigan Press.

Segura, Gerardo. 1996. *Forestry in Mexico*. Unpublished Report. Montreal: North American Commission for Environmental Cooperation.

SEMARNAP. 1996. *Programa de Areas Naturales Protegidas de Mexico, 1995-2000*. Mexico: Secretaría de Medio Ambiente, Recursos Naturales y Pesca (SEMARNAP), Instituto Nacional de Ecología.

Shiva, Vandana. 1993. "International Controversy over Sustainable Forestry." Pages 75-86 in *Green Globe Yearbook 1993*. New York: Oxford University Press.

Silva, Eduardo. 1994. "Thinking Politically about Sustainable Development in the Tropical Forests of Latin America." *Development and Change*, 25: 697-721.

Simonian, Lane. 1995. *Defending the Land of the Jaguar*. Austin: University of Texas Press.

S.R.H. 1976. *Estudio Integral de la Selva Lacandona*. Comisión del Río Grijalva, Subdirección de Estudios y Proyectos. Unpublished report.

Stepputat, Finn. 1989. *Self-Sufficiency and Exile in Mexico*. Discussion Paper 9: United Nations Research Institute for Social Development.

Stonich, Susan C. 1993. *I Am Destroying the Land: The Political Ecology of Poverty and Environmental Destruction in Honduras*. Boulder: Westview Press.

Stonich, Susan C., and Billie R. DeWalt. 1996. "The Political Ecology of Deforestation in Honduras." Pages 187-215 in L. E. Sponsel, T. N. Headland, and R. C. Bailey, eds., *Tropical Deforestation: The Human Dimension*. New York: Columbia University Press.

Thiesenhusen, William C. 1989. "Introduction: Searching for Agrarian Reform in Latin America." Pages 1 - 41 in W.C. Thiesenhusen, ed., *Searching for Reform in Latin America*. Boston: Unwin Hyman.

Thomspon, J.Eric S. 1970. *Maya History and Religion*. Norman, OK: University of Oklahoma Press.

Toledo, Alejandro, Arturo Nuñez, and Hector Ferreira. no date. *Como Destruir el Paraiso: El Desastre Ecologico del Sureste*. Mexico City: Centro de Ecodesarrollo.

Townsend, Janet G. 1996. "Pioneer Women and the Destruction of the Rainforests." Pages 108-114 in H. Collinson, ed., *Green Guerrillas: Environmental Conflicts and Initiatives in Latin America and the Caribbean*. London: Latin America Bureau.

Townsend, Janet, Ursula Arrevillaga, Jennie Bain, Socorro Cancino, Susan F. Frenk, Silvana Pacheco, and Elia Pérez. 1995. *Women's Voices from the Rainforest*. London: Routledge.

Trejo Delarbre, Raul, ed. 1994. *Chiapas: La Guerra de las Ideas*. Mexico: Editorial Diana.

Union de Uniones Ejidales y Grupos Campesinos Solidarios de Chiapas. 1983. "Nuestra Lucha por la Tierra en la Selva Lacandona: Balance de una Acción Campesina con Apoyo Obrero." *Textual, Analisis del Medio Rural (UACH)*, 4(13):150-161.

Vandermeer, John, and Ivette Perfecto. 1995. *Breakfast of Biodiversity: The Truth About Rain Forest Destruction*. Oakland, CA: The Institute for Food and Development Policy.

Vásquez Sánchez, Miguel Angel. 1992. "La Reserva de la Biósfera Montes Azules: Antecedentes." Pages 19-38 in M. A. Vásquez Sánchez and M. A. Ramos Olmos eds., *Reserva de la Biósfera Montes Azules, Selva Lacandona: Investigación para su Conservación*. San Cristóbal de las Casas, Chiapas: Publ. Esp. Ecosfera.

Vásquez Sánchez, Miguel Angel, and Mario A. Ramos Olmos, eds. 1992. *Reserva de la Biósfera Montes Azules, Selva Lacandona: Investigación para su Conservaction*. San Cristóbal de las Casas, Chiapas: Publ. Esp. Ecosfera.

Villafuerte Solís, Daniel, María del Carmen García Aguilar, and Salvador Meza Díaz. 1993. *Ganaderización: Deforestación en el Tropico Mexicano y sus Expresiones en el Estado de Chiapas*. Report to CINVESTAV-PROAFT (SARH), April 1993, Mexico. Available on Internet: http://www.ucr.edu/pril/peten/images/proaft/chiapas.html.

Villagran, Felipe. 1995. "Exposition on the Forestal Policies." *Letter to Dr. Oscar González Rodriguez, Subsecretary of Natural Resources, SEMARNAP*, April 18, 1995. Nature Net: http://bioc09.uthscsq.edu/natnet/archive/n1/9504/02 73.html.

Wasserstrom, Robert. 1983. *Class and Society in Central Chiapas*. Berkeley: University of California Press.

Weinberg, Bill. 1991. *War on the Land: Ecology and Politics in Central America*. London: Zed Books, Ltd.

Wellhausen, Edward J. 1976. "The Agriculture of Mexico." *Scientific American*, 235(3):128-150.

Williams, Michael. 1994. "Forests and Tree Cover." Pages 97-124 in W. B. Meyer and B. L. Turner II, eds., *Changes in Land Use and Land Cover: A Global Perspective*. Cambridge: Cambridge University Press.

Woodward, Laura Lynn, and Ralph Lee Woodward, Jr. 1985. Trudi Blom and the Lacandon Rain Forest. *Environmental Review*, 9(3):226-236.

World Commission on Environment and Development. 1987. *Our Common Future*. New York: Oxford University Press.

Yates, Paul Lamartine. 1981. *Mexico's Agricultural Dilemma*. Tucson: University of Arizona Press.

Referenced Web Sites

Conservation International:
  http://www.conservation.org/
  http://www.conservation.org/news/pressrel/mexswap.htm
  http://www.conservation.org/web/fieldact/regions/mcareg/Lacandon.htm
  http://www.conservation.org/web/fieldact/C-C_PROG/Aware/prvsectr.htm

La Jornada en Internet:
  http://www.sccs.swarthmore.edu/~justin/jornada

Man and the Biosphere Program (MAB):
  http://www.mabnet.org/brprogram/reserves.html

Montes Azules:
  http://www.ma.com.mx/montes.html

Programa de Desarrollo Sustentable (PRODERS)
  http://hp.fciencias.unam.mx/proders

Zapatista Communiqués:
  http://www.ezln.org/communiques/html

Index